**Teacher Edition**

# Eureka Math
# Grade 3
# Module 6

Special thanks go to the Gordon A. Cain Center and to the Department of Mathematics at Louisiana State University for their support in the development of *Eureka Math*.

For a free *Eureka Math* Teacher Resource Pack, Parent Tip Sheets,  and more please visit www.Eureka.tools

## *Eureka Math: A Story of Units* Contributors

Katrina Abdussalaam, Curriculum Writer
Tiah Alphonso, Program Manager—Curriculum Production
Kelly Alsup, Lead Writer / Editor, Grade 4
Catriona Anderson, Program Manager—Implementation Support
Debbie Andorka-Aceves, Curriculum Writer
Eric Angel, Curriculum Writer
Leslie Arceneaux, Lead Writer / Editor, Grade 5
Kate McGill Austin, Lead Writer / Editor, Grades PreK–K
Adam Baker, Lead Writer / Editor, Grade 5
Scott Baldridge, Lead Mathematician and Lead Curriculum Writer
Beth Barnes, Curriculum Writer
Bonnie Bergstresser, Math Auditor
Bill Davidson, Fluency Specialist
Jill Diniz, Program Director
Nancy Diorio, Curriculum Writer
Nancy Doorey, Assessment Advisor
Lacy Endo-Peery, Lead Writer / Editor, Grades PreK–K
Ana Estela, Curriculum Writer
Lessa Faltermann, Math Auditor
Janice Fan, Curriculum Writer
Ellen Fort, Math Auditor
Peggy Golden, Curriculum Writer
Maria Gomes, Pre-Kindergarten Practitioner
Pam Goodner, Curriculum Writer
Greg Gorman, Curriculum Writer
Melanie Gutierrez, Curriculum Writer
Bob Hollister, Math Auditor
Kelley Isinger, Curriculum Writer
Nuhad Jamal, Curriculum Writer
Mary Jones, Lead Writer / Editor, Grade 4
Halle Kananak, Curriculum Writer
Susan Lee, Lead Writer / Editor, Grade 3
Jennifer Loftin, Program Manager—Professional Development
Soo Jin Lu, Curriculum Writer
Nell McAnelly, Project Director

# Mathematics Curriculum

Table of Contents

# GRADE 3 • MODULE 6

## Collecting and Displaying Data

Grade 3 • Module 6

# Collecting and Displaying Data

## OVERVIEW

This 10-day module builds on Grade 2 concepts about data, graphing, and line plots. Topic A begins with a lesson in which students generate categorical data, organize it, and then represent it in a variety of forms. Drawing on Grade 2 knowledge, students might initially use tally marks, tables, or graphs with one-to-one correspondence. By the end of the lesson, they show data in tape diagrams where units are equal groups with a value greater than 1. In the next two lessons, students rotate the tape diagrams vertically so that the tapes become the units or bars of scaled graphs (**3.MD.3**). Students understand picture and bar graphs as vertical representations of tape diagrams and apply well-practiced skip-counting and multiplication strategies to analyze them. In Lesson 4, students synthesize and apply learning from Topic A to solve one- and two-step problems. Through problem solving, opportunities naturally surface for students to make observations, analyze, and answer questions such as, "How many more?" or "How many less?" (**3.MD.3**).

In Topic B, students learn that intervals do not have to be whole numbers but can have fractional values that facilitate recording measurement data with greater precision. In Lesson 5, they generate a six-inch ruler marked in whole-inch, half-inch, and quarter-inch increments, using the Module 5 concept of partitioning a whole into parts. This creates a conceptual link between measurement and recent learning about fractions. Students then use the rulers to measure the lengths of precut straws and record their findings to generate measurement data (**3.MD.4**).

Lesson 6 reintroduces line plots as a tool for displaying measurement data. Although familiar from Grade 2, line plots in Grade 3 have the added complexity of including fractions on the number line (**2.MD.9**, **3.MD.4**). In this lesson, students interpret scales involving whole, half, and quarter units in order to analyze data. This experience lays the foundation for them to create their own line plots in Lessons 7 and 8. To draw line plots, students learn to choose appropriate intervals within which to display a particular set of data. For example, to show measurements of classmates' heights, students might notice that their data fall within the range of 45 to 55 inches and then construct a line plot with the corresponding interval.

Students end the module by applying learning from Lessons 1–8 to problem solving. They work with a mixture of scaled picture graphs, bar graphs, and line plots to problem solve using both categorical and measurement data (**3.MD.3**, **3.MD.4**).

### Notes on Pacing for Differentiation

If pacing is a challenge, consider the following modifications and omissions.

Omit Lesson 9, a problem solving lesson involving categorical and measurement data. Be sure to embed problem solving practice with both types of data into prior lessons.

## Distribution of Instructional Minutes

This diagram represents a suggested distribution of instructional minutes based on the emphasis of particular lesson components in different lessons throughout the module.

■ Fluency Practice
▢ Concept Development
■ Application Problems
■ Student Debrief

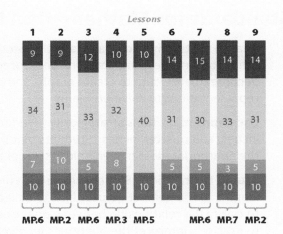

MP = Mathematical Practice

# Focus Grade Level Standards

### Represent and interpret data.

**3.MD.3**    Draw a scaled picture graph and a scaled bar graph to represent a data set with several categories. Solve one- and two-step "how many more" and "how many less" problems using information presented in scaled bar graphs. *For example, draw a bar graph in which each square in the bar graph might represent 5 pets.*

**3.MD.4**    Generate measurement data by measuring lengths using rulers marked with halves and fourths of an inch. Show the data by making a line plot, where the horizontal scale is marked off in appropriate units—whole numbers, halves, or quarters.

# Foundational Standards

**2.MD.5**    Use addition and subtraction within 100 to solve word problems involving lengths that are given in the same units, e.g., by using drawings (such as drawings of rulers) and equations with a symbol for the unknown number to represent the problem.

**2.MD.6**    Represent whole numbers as lengths from 0 on a number line diagram with equally spaced points corresponding to the numbers 0, 1, 2, ..., and represent whole-number sums and differences within 100 on a number line diagram.

**2.MD.9**    Generate measurement data by measuring lengths of several objects to the nearest whole unit, or by making repeated measurements of the same object. Show the measurements by making a line plot, where the horizontal scale is marked off in whole-number units.

**2.MD.10**    Draw a picture graph and a bar graph (with single-unit scale) to represent a data set with up to four categories. Solve simple put-together, take-apart, and compare problems[1] using information presented in a bar graph.

---

[1]See Glossary, Table 1.

# Focus Standards for Mathematical Practice

**MP.2**  **Reason abstractly and quantitatively.** Students work with data in the context of science and other content areas and interpret measurement data using line plots. Students decontextualize data to create graphs and then contextualize as they analyze their representations to solve problems.

**MP.5**  **Use appropriate tools strategically.** Students create and use rulers marked in inches, half inches, and quarter inches. Students plot measurement data on a line plot and reason about the appropriateness of a line plot as a tool to display fractional measurements.

**MP.6**  **Attend to precision.** Students generate rulers using precise measurements and then measure lengths to the nearest quarter inch to collect and record data. Students label axes on graphs to clarify the relationship between quantities and units and attend to the scale on the graph to precisely interpret the quantities involved.

**MP.7**  **Look for and make use of structure.** Students use an auxiliary line to create equally spaced increments on a six-inch strip, which is familiar from the previous module. Students look for trends in data to help solve problems and draw conclusions about the data.

# Overview of Module Topics and Lesson Objectives

| Standards | Topics and Objectives | | Days |
|---|---|---|---|
| **3.MD.3** | A | **Generate and Analyze Categorical Data** | 4 |
| | | Lesson 1:  Generate and organize data. | |
| | | Lesson 2:  Rotate tape diagrams vertically. | |
| | | Lesson 3:  Create scaled bar graphs. | |
| | | Lesson 4:  Solve one- and two-step problems involving graphs. | |
| **3.MD.4** | B | **Generate and Analyze Measurement Data** | 5 |
| | | Lesson 5:  Create ruler with 1-inch, $\frac{1}{2}$-inch, and $\frac{1}{4}$-inch intervals, and generate measurement data. | |
| | | Lesson 6:  Interpret measurement data from various line plots. | |
| | | Lessons 7–8:  Represent measurement data with line plots. | |
| | | Lesson 9:  Analyze data to problem solve. | |
| | | End-of-Module Assessment: Topics A–B  (assessment ½ day, return ¼ day, remediation or further applications ¼ day) | 1 |
| **Total Number of Instructional Days** | | | 10 |

# Terminology

## New or Recently Introduced Terms

- Frequent (most common measurement on a line plot)
- Key (notation on a graph explaining the value of a unit)
- Measurement data (e.g., length measurements of a collection of pencils)
- Scaled graphs (bar or picture graph in which the scale uses units with a value greater than 1)

## Familiar Terms and Symbols[2]

- Bar graph (graph generated from categorical data with bars to represent a quantity)
- Data (information)
- Fraction (numerical quantity that is not a whole number, e.g., $\frac{1}{3}$)
- Line plot (display of data on a horizontal line)
- Picture graph (graph generated from categorical data with graphics to represent a quantity)
- Scale (a number line used to indicate the various quantities represented in a bar graph)
- Survey (collecting data by asking a question and recording responses)

# Suggested Tools and Representations

- Bar graph
- Grid paper
- Line plot
- Picture graph
- Rulers (measuring in inches, half inches, and quarter inches)
- Sentence strips
- Tape diagram

---

[2]These are terms and symbols students have seen previously.

# Scaffolds[3]

The scaffolds integrated into *A Story of Units* give alternatives for how students access information as well as express and demonstrate their learning. Strategically placed margin notes are provided within each lesson elaborating on the use of specific scaffolds at applicable times. They address many needs presented by English language learners, students with disabilities, students performing above grade level, and students performing below grade level. Many of the suggestions are organized by Universal Design for Learning (UDL) principles and are applicable to more than one population. To read more about the approach to differentiated instruction in *A Story of Units,* please refer to "How to Implement *A Story of Units.*"

## Assessment Summary

| Type | Administered | Format | Standards Addressed |
|------|-------------|--------|--------------------|
| End-of-Module Assessment Task | After Topic B | Constructed response with rubric | 3.MD.3<br>3.MD.4 |

*\*Because this module is short, there is no Mid-Module Assessment. Module 6 should normally be completed just prior to the state assessment. This may not be true, however, depending on variations in pacing. In the case that it is not true, be aware that 3.MD.3 (addressed in Topic A) is a pretest standard, while 3.MD.4 (addressed in Topic B) is a post-test standard.*

---

[3]Students with disabilities may require Braille, large print, audio, or special digital files. Please visit the website www.p12.nysed.gov/specialed/aim for specific information on how to obtain student materials that satisfy the National Instructional Materials Accessibility Standard (NIMAS) format.

EUREKA MATH™

# Mathematics Curriculum

## Topic A

# Generate and Analyze Categorical Data

## 3.MD.3

| Focus Standard: | 3.MD.3 | Draw a scaled picture graph and a scaled bar graph to represent a data set with several categories.  Solve one- and two-step "how many more" and "how many less" problems using information presented in scaled bar graphs. *For example, draw a bar graph in which each square in the bar graph might represent 5 pets.* |
|---|---|---|
| **Instructional Days:** | 4 | |
| **Coherence**   **-Links from:** | G2–M7 | Problem Solving with Length, Money, and Data |
| | G3–M1 | Properties of Multiplication and Division and Solving Problems with Units of 2–5 and 10 |
| **-Links to:** | G4–M2 | Unit Conversions and Problem Solving with Metric Measurement |
| | G4–M7 | Exploring Measurement with Multiplication |

Drawing on prior knowledge from Grade 2, students generate categorical data from community-building activities.  In Lesson 1, they organize the data and then represent them in a variety of ways (e.g., tally marks, graphs with one-to-one correspondence, or tables).  By the end of the lesson, students show data as picture graphs where each picture has a value greater than 1.

Students rotate tape diagrams vertically in Lesson 2.  These rotated tape diagrams with units of values other than 1 help transition students toward creating scaled bar graphs in Lesson 3.  Bar and picture graphs are introduced in Grade 2; however, Grade 3 adds the complexity that one unit—one picture or unit on the bar— can have a whole number value greater than 1.  Students practice familiar skip-counting and multiplication strategies with rotated tape diagrams to bridge understanding that these same strategies can be applied to problem solving with bar graphs.

In Lesson 3, students construct the scale on the vertical axis of a bar graph.  One rotated tape becomes one bar on the bar graph.  As with the unit of a tape diagram, one unit of a bar graph can have a value greater than 1.  Students create number lines with intervals appropriate to the data.

Lesson 4 provides an opportunity for students to analyze graphs and to solve more sophisticated one- and two-step problems, including comparison problems.  This work highlights Mathematical Practice 2 as students re-contextualize their numerical work to interpret its meaning as data.

| A Teaching Sequence Toward Mastery to Generate and Analyze Categorical Data |
| --- |

**Objective 1:** Generate and organize data.
(Lesson 1)

**Objective 2:** Rotate tape diagrams vertically.
(Lesson 2)

**Objective 3:** Create scaled bar graphs.
(Lesson 3)

**Objective 4:** Solve one- and two-step problems involving graphs.
(Lesson 4)

# Lesson 1

Objective: Generate and organize data.

## Suggested Lesson Structure

■ Fluency Practice          (9 minutes)
▨ Application Problem        (7 minutes)
▫ Concept Development        (34 minutes)
■ Student Debrief            (10 minutes)
   **Total Time**          **(60 minutes)**

## Fluency Practice  (9 minutes)

- Group Counting on a Vertical Number Line  **3.OA.1**      (3 minutes)
- Model Division with Tape Diagrams  **3.MD.4**              (6 minutes)

### Group Counting on a Vertical Number Line  (3 minutes)

Note:  Group counting reviews interpreting multiplication as repeated addition.

  T:  (Project a vertical number line partitioned into intervals of 6, as shown.  Cover the number line so that only the numbers 0 and 12 show.)  What is halfway between 0 and 12?

  S:  6.

  T:  (Write 6 on the first hash mark.)

Continue for the remaining hashes so that the number line shows increments of six to 60.

  T:  Let's count by sixes to 60.

Direct students to count forward and backward to 60, occasionally changing the direction of the count.  Repeat the process with the following possible suggestions:

-  Sevens to 70
-  Eights to 80
-  Nines to 90

### Model Division with Tape Diagrams  (6 minutes)

Materials:  (S) Personal white board

Note:  This fluency activity reviews using tape diagrams to model division.

  T:  (Project tape diagram with 6 as the whole.)  What is the value of the whole?

  S:  6.

**Lesson 1:**    Generate and organize data.

9

T:   (Partition the tape diagram into 2 equal parts.)  How many equal parts is 6 broken into?

S:   2 equal parts.

T:   Tell me a division equation to solve for the unknown group size.

S:   $6 \div 2 = 3$.

T:   (Beneath the diagram, write $6 \div 2 = 3$.)

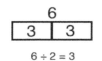

T:   On your personal white board, draw a rectangle with 8 as the whole.

S:   (Draw a rectangle with 8 as the whole.)

T:   Divide it into 2 equal parts, write a division equation to solve for the unknown, and label the value of the units.

S:   (Partition the rectangle into 2 equal parts, write $8 \div 2 = 4$, and label each unit with 4.)

Continue with the following possible suggestions, alternating between teacher drawings and student drawings:  $6 \div 3$, $8 \div 4$, $10 \div 5$, $10 \div 2$, $9 \div 3$, $12 \div 2$, $12 \div 3$, and $12 \div 4$.

## Application Problem  (7 minutes)

Damien folds a paper strip into 6 equal parts.  He shades 5 of the equal parts and then cuts off 2 shaded parts.  Explain your thinking about what fraction is unshaded.

Note:  This Application Problem provides an opportunity to review the concept of defining the whole from Module 5.  Some students may correctly argue that one-fourth is unshaded if they see the strip as a new whole partitioned into fourths.

Cut

$\frac{1}{6}$ of the paper strip is unshaded. After 2 sixths are cut, 3 sixths are still shaded and 1 sixth is unshaded.

## Concept Development  (34 minutes)

Materials:   (S) Problem Set, class list (preferably in alphabetical order, as shown to the right)

**Part 1:  Collect data.**

List the following five colors on the board:  green, yellow, red, blue, and orange.

T:   Today you will collect information, or data.  We will use a survey to find out each person's favorite color from one of the five colors listed on the board.  How can we keep track of our data in an organized way?  Turn and talk to your partner.

S:   We can write everyone's name with the person's favorite color next to it.
     → We can write each name and color code it with the person's favorite color.
     → We can put it in a chart.

Room 7 - Third Grade
Mrs. Lee / Mrs. Prescott
Anna
Caleb
Charlotte
Christopher
Emily
Hyun Soo
James
Jia
Jinhee
John
Joseph
Josh
Jung
Kathy
Kylie
Matthew
Miae
Noah
Steven
Susan
William

EUREKA
MATH™

T:  All of those ways work.  One efficient way to collect and organize our data is by recording it on a tally chart. (Draw a single vertical tally mark on the board.)  Each tally like the one I drew has a value of 1 student. Count with me.  (Draw tally marks as students count.)

S:  1 student, 2 students, 3 students, 4 students, 5 students.

T:  (Draw ||||.)  This is how 5 is represented with tally marks.  How might writing each fifth tally mark with a slash help you count your data easily and quickly?  Talk to your partner.

S:  It is bundling tally marks by fives.
    → We can bundle 2 fives as ten.

T:  (Pass out the Problem Set and class list.)  Find the chart on Problem 1 of your Problem Set (pictured to the right).  Take a minute now to choose your favorite color out of those listed on the chart. Record your favorite color with a tally mark on the chart, and cross your name off your class list.

T:  (Students record.)  Take six minutes to ask each of your classmates, "What is your favorite color?"  Record each classmate's answer with a tally mark next to his favorite color.  Once you are done with each person, cross the person's name off your class list to help you keep track of who you still need to ask.  Remember, you may not change your color throughout the survey.

S:  (Conduct the survey for about six minutes.)

T:  How many total students said green was their favorite color?

S:  (Say the number of students.)

T:  I am going to record it numerically on the board below the label *Green*.

Continue with the rest of the colors.

T:  This chart is another way to show the same information.

T:  Use mental math to find the total number of students surveyed.  Say the total at my signal.  (Signal.)

S:  22 students.

**NOTES ON VOCABULARY:**

Students are familiar with tally marks and tally charts from their work in Grades 1 and 2.  In Grades 1 and 2 they also used the word *table* to refer to these charts.

1.  "What is your favorite color?"  Survey the class to complete the tally chart below.

| Favorite Colors | |
| --- | --- |
| Color | Number of Students |
| Green | |
| Yellow | |
| Red | |
| Blue | |
| Orange | |

**NOTES ON MULTIPLE MEANS OF REPRESENTATION:**

Familiarize English language learners and others with common language used to discuss data, such as *most common, favorite, how many more,* and *how many less.*  Offer explanations in students' first language, if appropriate.  Guiding students to use the language to quickly ask questions about the tally chart at this point in the Concept Development prepares them for independent work on the Problem Set.

Example Board:

| Green | Yellow | Red | Blue | Orange |
| --- | --- | --- | --- | --- |
| 4 | 2 | 6 | 7 | 3 |

Total:  $4 + 2 + 6 + 7 + 3 = 22$

T:   Discuss your mental math with your partner for 30 seconds.

S:   I added 4 and then 2 to get 6.  Six and 6 is 12, and then I noticed I had 10 left.  Twelve and 10 is 22.
     → I made 2 tens—6 plus 4 and 7 plus 3—and then, I added 2 more.

**Part 2:  Construct a picture graph from the data.**

T:   Using pictures or a picture graph, let's graph the data
     we collected.  Read the directions for Problem 3 on
     your Problem Set (pictured to the right).  (Pause for
     students to read.)  Find the **key**, which tells you the
     value of a unit, on each picture graph.  (Pause for
     students to locate the keys.)  What is different about
     the keys on these two picture graphs?

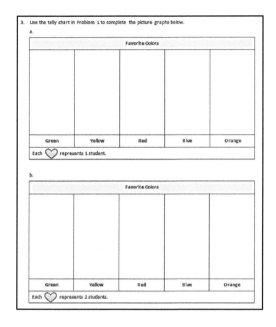

S:   In Problem 3(a), one heart represents 1 student, but in
     Problem 3(b), one heart represents 2 students.

T:   Good observations!  Talk to a partner:  How would you
     represent 4 students in Problems 3(a) and 3(b)?

S:   In Problem 3(a), I would draw 4 hearts.  → In Problem
     3(b), I would only draw 2 hearts because the value of
     each heart is 2 students.

T:   (Draw 🤍 🤍 🤍.)  Each heart represents
     2 students, like in Problem 3(b).  What is the value of
     this picture?

S:   6 students.

T:   Write a multiplication sentence to represent the value
     of my picture, where the number of hearts is the
     number of groups, and the number of students is the
     size of each group.

S:   (Write $3 \times 2 = 6$.)

**MP.6**  T:   Turn and talk:  How can we use the hearts to represent
     an odd number like 5?

S:   We can draw 3 hearts and then cross off a part of
     1 heart to represent 5.  → We can show half of a heart
     to represent 1 student.

T:   What is the value of half of 1 heart?

S:   1 student.

T:   I can estimate to erase half of 1 heart.
     (Erase half of 1 heart to show 🤍 🤍 (.)  Now, my
     picture represents a value of 5.

T:   Begin filling out the picture graphs in Problem 3.
     Represent your tally chart data as hearts and half-
     hearts to make your picture graphs.

**NOTES ON
MULTIPLE MEANS
OF ACTION AND
EXPRESSION:**

Precise sketching of hearts drawn in
the picture graph of Problem 3 may
prove challenging for students working
below grade level and others.  The task
of completing the picture graph may be
eased by providing pre-cut hearts and
half-hearts that can be glued.
Alternatively, offer the option to draw
a more accessible picture, such as a
square.  If students choose a different
picture, they need to be sure to change
the key in order to reflect their choice.

## Problem Set  (10 minutes)

Students should do their personal best to complete Problems 2 and 4 within the allotted 10 minutes.  Some problems do not specify a method for solving.  This is an intentional reduction of scaffolding that invokes MP.5, Use Appropriate Tools Strategically.  Students should solve these problems using the RDW approach used for Application Problems.

For some classes, it may be appropriate to modify the assignment by specifying which problems students should work on first.  With this option, let the careful sequencing of the Problem Set guide the selections so that problems continue to be scaffolded.  Balance word problems with other problem types to ensure a range of practice.  Assign incomplete problems for homework or at another time during the day.

## Student Debrief  (10 minutes)

**Lesson Objective:**  Generate and organize data.

The Student Debrief is intended to invite reflection and active processing of the total lesson experience.

Invite students to review their solutions for the Problem Set.  They should check work by comparing answers with a partner before going over answers as a class.  Look for misconceptions or misunderstandings that can be addressed in the Debrief.  Guide students in a conversation to debrief the Problem Set and process the lesson.

Any combination of the questions below may be used to lead the discussion.

- Compare the data in the picture graphs in Problems 3(a) and 3(b).
- Share answers to Problems 4(c) and 4(d).  What would Problem 4(d) look like as a multiplication sentence?
- Compare picture graphs with tally charts.  What makes each one useful?  What are the limitations of each?
- Why is it important to use the **key** to understand the value of a unit in a picture graph?

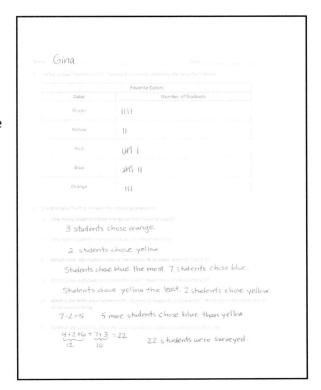

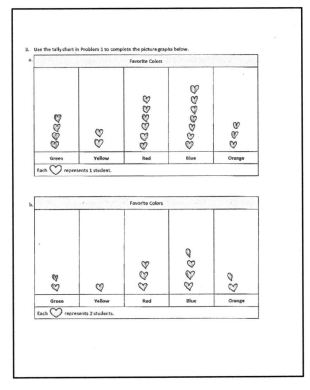

- What math vocabulary did we use today to talk about recording and gathering information? (*data*, *survey*)

## Exit Ticket  (3 minutes)

After the Student Debrief, instruct students to complete the Exit Ticket.  A review of their work will help with assessing students' understanding of the concepts that were presented in today's lesson and planning more effectively for future lessons.  The questions may be read aloud to the students.

4. Use the picture graph in Problem 3(b) to answer the following questions.

   a. What does each ♡ represent?

      Each ♡ represents 2 students.

   b. Draw a picture and write a number sentence to show how to represent 3 students in your picture graph.

      ♡ ♡    2 + 1 = 3
      2   1

   c. How many students does ♡ ♡ ♡ ♡ ♡ ♡ ♡ represent? Write a number sentence to show how you know.

      7 × 2 = 14
      It represents 14 students.

   d. How many more ♡ did you draw for the color that students chose the most than for the color that students chose the least? Write a number sentence to show the difference between the number of votes for the color that students chose the most and the color that students chose the least.

      I drew 2½ more hearts for blue than for yellow
      7 - 2 = 5
      5 more students chose blue than yellow

**Lesson 1:**      Generate and organize data.

EUREKA MATH™

Name _____ Date _____

1. "What is your favorite color?"  Survey the class to complete the tally chart below.

| Favorite Colors | |
|---|---|
| **Color** | **Number of Students** |
| **Green** | |
| **Yellow** | |
| **Red** | |
| **Blue** | |
| **Orange** | |

2. Use the tally chart to answer the following questions.

   a. How many students chose orange as their favorite color?

   b. How many students chose yellow as their favorite color?

   c. Which color did students choose the most?  How many students chose it?

   d. Which color did students choose the least?  How many students chose it?

   e. What is the difference between the number of students in parts (c) and (d)?  Write a number sentence to show your thinking.

   f. Write an equation to show the total number of students surveyed on this chart.

3. Use the tally chart in Problem 1 to complete the picture graphs below.

   a.

| Favorite Colors | | | | |
|---|---|---|---|---|
| | | | | |
| **Green** | **Yellow** | **Red** | **Blue** | **Orange** |

   Each ♡ represents 1 student.

   b.

| Favorite Colors | | | | |
|---|---|---|---|---|
| | | | | |
| **Green** | **Yellow** | **Red** | **Blue** | **Orange** |

   Each ♡ represents 2 students.

Lesson 1:     Generate and organize data.

**EUREKA MATH**

4.  Use the picture graph in Problem 3(b) to answer the following questions.

    a.  What does each ♡ represent?

    b.  Draw a picture and write a number sentence to show how to represent 3 students in your picture graph.

    c.  How many students does ♡ ♡ ♡ ♡ ♡ ♡ ♡ represent?  Write a number sentence to show how you know.

    d.  How many more ♡ did you draw for the color that students chose the most than for the color that students chose the least?  Write a number sentence to show the difference between the number of votes for the color that students chose the most and the color that students chose the least.

Name _____    Date _____

The picture graph below shows data from a survey of students' favorite sports.

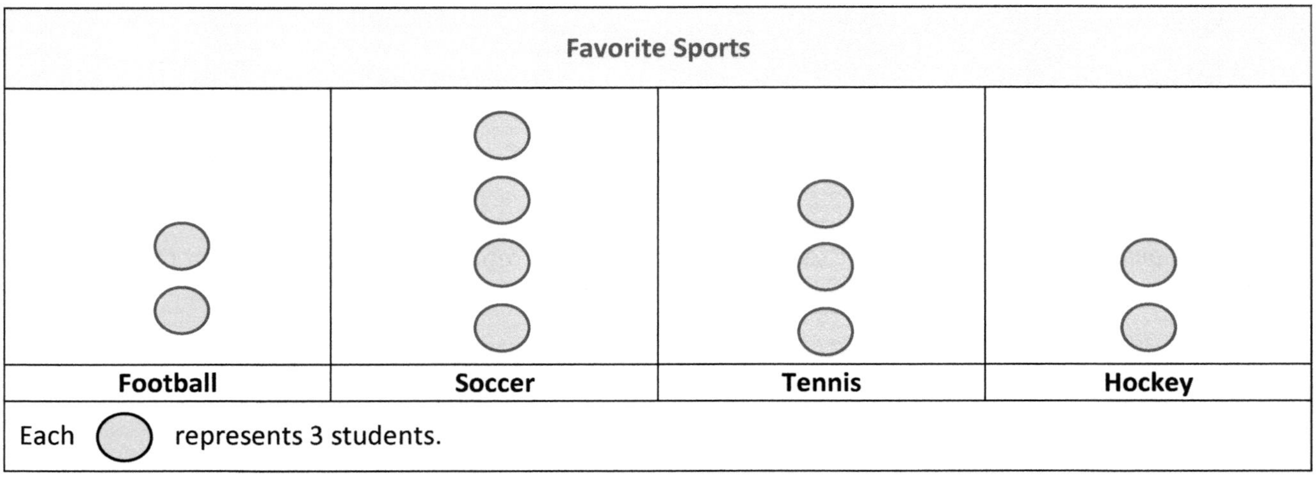

| Favorite Sports | | | |
|---|---|---|---|
| ◯ ◯ | ◯ ◯ ◯ ◯ | ◯ ◯ ◯ | ◯ ◯ |
| **Football** | **Soccer** | **Tennis** | **Hockey** |

Each ◯ represents 3 students.

a. The same number of students picked _____ and _____ as their favorite sport.

b. How many students picked tennis as their favorite sport?

c. How many more students picked soccer than tennis? Use a number sentence to show your thinking.

d. How many total students were surveyed?

**Lesson 1:**    Generate and organize data.

**EUREKA MATH**

Name _____   Date _____

1.  The tally chart below shows a survey of students' favorite pets.  Each tally mark represents 1 student.

| Favorite Pets | |
|---|---|
| Pets | Number of Pets |
| Cats | ~~////~~ / |
| Turtles | //// |
| Fish | // |
| Dogs | ~~////~~  /// |
| Lizards | // |

The chart shows a total of _____ students.

2.  Use the tally chart in Problem 1 to complete the picture graph below.  The first one has been done for you.

| Favorite Pets | | | | |
|---|---|---|---|---|
| ◯ ◯ ◯ ◯ ◯ ◯ | | | | |
| Cats | Turtles | Fish | Dogs | Lizards |

Each ◯ represents 1 student.

a.  The same number of students picked _____ and _____ as their favorite pet.

b.  How many students picked dogs as their favorite pet?

c.  How many more students chose cats than turtles as their favorite pet?

3. Use the tally chart in Problem 1 to complete the picture graph below.

| Favorite Pets | | | | |
|---|---|---|---|---|
| | | | | |
| Cats | Turtles | Fish | Dogs | Lizards |

Each ☐ represents 2 students.

a. What does each ☐ represent?

b. How many students does ☐ ☐ ☐ ☐ ☐ represent? Write a number sentence to show how you know.

c. How many more ☐ did you draw for dogs than for fish? Write a number sentence to show how many more students chose dogs than fish.

Lesson 1:  Generate and organize data.

©2015 Great Minds. eureka-math.org
G3-M6-TE-B6-1.3.1-01.2016

EUREKA MATH™

# Lesson 2

Objective:  Rotate tape diagrams vertically.

## Suggested Lesson Structure

■ Fluency Practice          (9 minutes)
▨ Application Problem        (10 minutes)
▢ Concept Development        (31 minutes)
■ Student Debrief           (10 minutes)

   **Total Time**            **(60 minutes)**

## Fluency Practice  (9 minutes)

- Group Counting on a Vertical Number Line  **3.OA.1**      (3 minutes)
- Read Tape Diagrams  **3.MD.4**                            (6 minutes)

### Group Counting on a Vertical Number Line  (3 minutes)

Note:  Group counting reviews interpreting multiplication as repeated addition.

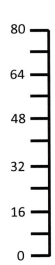

   T:  (Project a vertical number line partitioned into intervals of 8, as shown. Cover the number line so that only the numbers 0 and 16 show.)  What is halfway between 0 and 16?

   S:  8.

   T:  (Write 8 on the first hash mark.)

Continue for the remaining hashes so that the number line shows increments of eight to 80.

   T:  Let's count by eights to 80.

Direct students to count forward and backward to 80, occasionally changing the direction of the count.  Repeat the process using the following possible suggestions:

- Sixes to 60
- Sevens to 70
- Nines to 90

EUREKA
MATH™

**Read Tape Diagrams  (6 minutes)**

Materials:   (S) Personal white board

Note:  This fluency activity reviews the relationship between the value of each unit in a tape diagram and the total value of the tape diagram.  It also reviews comparing tape diagrams in preparation for today's lesson.

T:   (Project a tape diagram with 7 units.)  Each unit in the tape diagram has a value of 4.  Write a multiplication sentence that represents the total value of the tape diagram.

S:   (Write 7 × 4 = 28.)

T:   What is the total value of the tape diagram?

S:   28.

Use the same tape diagram.  Repeat the process with the following suggested values for the units:  6, 3, 9, 7, and 8.

T:   (Project the tape diagrams as shown.)
Diagrams A and B?

S:   8.

What is the value of each unit in Tape

A:  | 8 | 8 | 8 | 8 |

B:  | 8 | 8 | 8 | 8 | 8 | 8 | 8 |

T:   Write a multiplication sentence that represents the total value of Tape Diagram A.

S:   (Write 4 × 8 = 32.)

T:   Write a multiplication sentence that represents the total value of Tape Diagram B.

S:   (Write 7 × 8 = 56.)

Continue with the following possible questions:

- What is the total value of both tape diagrams?
- How many more units of 8 are in Tape Diagram B than in Tape Diagram A?
- What is the difference in value between the 2 tape diagrams?

## Application Problem  (10 minutes)

Reisha played in three basketball games.  She scored 12 points in Game 1, 8 points in Game 2, and 16 points in Game 3.  Each basket that she made was worth 2 points.  She uses tape diagrams with a unit size of 2 to represent the points she scored in each game.  How many total units of 2 does it take to represent the points she scored in all three games?

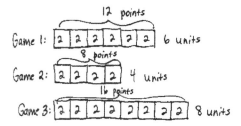

Total units of 2:  6 + 4 + 8 = 18
It will take 18 total units of 2 to represent the points scored in all 3 games.

Note: This problem reviews building tape diagrams with a unit size larger than 1 in anticipation of students using this same skill in the Concept Development. Ask students to solve this problem on personal white boards so that they can easily modify their work as they use it in the Concept Development. Invite students to discuss what the total number of units represents in relation to the three basketball games (18 total units of 2 is equal to 18 total baskets scored).

## Concept Development (31 minutes)

Materials:  (S) Tape diagrams from Application Problem, personal white board

**Problem 1:  Rotate tape diagrams to make vertical tape diagrams with units of 2.**

T:  Turn your personal white board so your tape diagrams are vertical like mine. (Model.) Erase the brackets and the labels for the number of units and the points. How are these vertical tape diagrams similar to the picture graphs you made yesterday?

S:  They both show us data. → Each unit on the vertical tape diagrams represents 2 points. → The pictures on the picture graph had a value greater than 1, and so does the unit in the vertical tape diagram.

T:  How are the vertical tape diagrams different from the picture graphs?

S:  The units are connected in the vertical tape diagrams. The pictures were separate in the picture graphs. → The units in the vertical tape diagrams are labeled, but in our picture graphs the value of the unit was shown on the bottom of the graph.

T:  Nice observations. Put your finger on the tape that shows data about Game 1. Now, write a multiplication equation to show the value of Game 1's tape.

S:  (Write $6 \times 2 = 12$.)

T:  What is the value of Game 1's tape?

S:  12 points!

T:  How did you know that the unit is points?

S:  The Application Problem says Reisha scores 12 points in Game 1.

T:  Let's write a title on our vertical tape diagrams to help others understand our data. What do the data on the vertical tape diagrams show us?

S:  The points Reisha scores in three basketball games.

T:  Write *Points Reisha Scores* for your title. (Model appropriate placement of the title.)

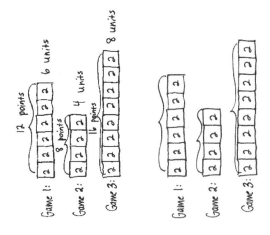

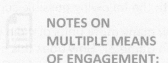

**NOTES ON MULTIPLE MEANS OF ENGAGEMENT:**

Students working above grade level and others may use parentheses and variables in their equations that represent the total points scored in all three games. Celebrate all true expressions, particularly those that apply the distributive property.

Students working below grade level and others may benefit from more scaffolded instruction for constructing and solving equations for three addends (number of units) and the total points.

**Problem 2:  Draw vertical tape diagrams with units of 4.**

T:  Suppose each unit has a value of 4 points instead of 2 points.  Talk to a partner.  How many units should I draw to represent Reisha's points in Game 1?  How do you know?

S:  Three units because she scored 12 points in Game 1, and 3 units of 4 points equals 12 points.  → Three units because 3 × 4 = 12 or 12 ÷ 4 = 3.  → Three units.  The value of each unit is twice as much.  Before we drew 6 units of 2, so now we draw half as many.  Each new unit has the value of two old units.

T:  Draw the 3 units vertically, and label each unit 4.  (Model.)  What label do we need for this tape?

S:  Game 1.

Continue the process for Games 2 and 3.

T:  How many total units of 4 does it take to represent the points Reisha scored in all three games?

S:  9 units!

T:  How does this compare to the total units of 2 it takes to represent Reisha's total points?

S:  It takes half as many total units when we used units of 4.

T:  Why does it take fewer units when you use units of 4?

S:  The units are bigger.  → The units represent a larger amount.

T:  How can you use vertical tape diagrams to write a multiplication sentence to represent Reisha's total points in all three games?

S:  Multiply the total number of units times the value of each unit.  → We can multiply 9 times 4.

T:  Write a multiplication number sentence to show the total points Reisha scored in all three games.

S:  (Write 9 × 4 = 36.)

T:  How many points did Reisha score in all three games?

S:  36 points!

Continue with the following possible suggestions:

**MP.2**

- How many more units of 4 did you draw for Game 1 than Game 2?  How does this help you find how many more points Reisha scored in Game 1 than in Game 2?

- Suppose Reisha scored 4 fewer points in Game 3.  How many units of 4 do you need to erase from Game 3's tape to show the new points?

- Reisha scores 21 points in a fourth game.  Can you use units of 4 to represent the points Reisha scores in Game 4 on a vertical tape diagram?

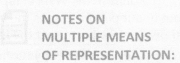

NOTES ON
MULTIPLE MEANS
OF REPRESENTATION:

In Problem 1 of the Problem Set, some students with perceptual challenges may have difficulty tracking rows of stamps as they count.  Have students place a straightedge below each row as they count by fours.  Students working below grade level may benefit from a fluency drill that reviews the fours group count.

EUREKA
MATH™

## Problem Set  (10 minutes)

Students should do their personal best to complete the Problem Set within the allotted 10 minutes.  For some classes, it may be appropriate to modify the assignment by specifying which problems they work on first.  Some problems do not specify a method for solving.  Students should solve these problems using the RDW approach used for Application Problems.

## Student Debrief  (10 minutes)

**Lesson Objective:**  Rotate tape diagrams vertically.

The Student Debrief is intended to invite reflection and active processing of the total lesson experience.

Invite students to review their solutions for the Problem Set.  They should check work by comparing answers with a partner before going over answers as a class.  Look for misconceptions or misunderstandings that can be addressed in the Debrief.  Guide students in a conversation to debrief the Problem Set and process the lesson.

Any combination of the questions below may be used to lead the discussion.

- How does multiplication help you interpret the vertical tape diagrams on the Problem Set?
- Could you display the data in Problem 1 in a vertical tape diagram with units of 6?  Why or why not?
- If the value of the unit for your vertical tape diagrams in Problem 1 was 2 instead of 4, how would the number of units change?
- In what ways do vertical tape diagrams relate to picture graphs?
- How did today's Application Problem relate to our new learning?
- In what ways did the Fluency Practice prepare you for today's lesson?

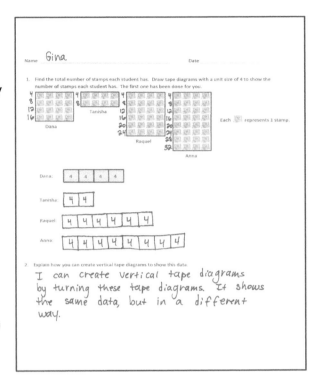

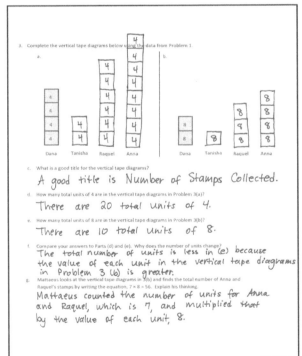

## Exit Ticket (3 minutes)

After the Student Debrief, instruct students to complete the Exit Ticket. A review of their work will help with assessing students' understanding of the concepts that were presented in today's lesson and planning more effectively for future lessons. The questions may be read aloud to the students.

**Lesson 2:** Rotate tape diagrams vertically.

©2015 Great Minds. eureka-math.org
G3-M6-TE-B6-1.3.1-01.2016

EUREKA
MATH™

Name _____ Date _____

1. Find the total number of stamps each student has. Draw tape diagrams with a unit size of 4 to show the number of stamps each student has. The first one has been done for you.

Dana

Tanisha

Raquel

Anna

Each ▨ represents 1 stamp.

Dana: | 4 | 4 | 4 | 4 |

Tanisha:

Raquel:

Anna:

2. Explain how you can create vertical tape diagrams to show this data.

3. Complete the vertical tape diagrams below using the data from Problem 1.

a.

| 4 |
|---|
| 4 |
| 4 |
| 4 |

Dana    Tanisha    Raquel    Anna

b.

| 8 |
|---|
| 8 |

Dana    Tanisha    Raquel    Anna

c. What is a good title for the vertical tape diagrams?

d. How many total units of 4 are in the vertical tape diagrams in Problem 3(a)?

e. How many total units of 8 are in the vertical tape diagrams in Problem 3(b)?

f. Compare your answers to parts (d) and (e). Why does the number of units change?

g. Mattaeus looks at the vertical tape diagrams in Problem 3(b) and finds the total number of Anna's and Raquel's stamps by writing the equation 7 × 8 = 56. Explain his thinking.

Lesson 2:    Rotate tape diagrams vertically.

EUREKA
MATH™

Name _____   Date _____

The chart below shows a survey of the book club's favorite type of book.

| Book Club's Favorite Type of Book | |
| --- | --- |
| Type of Book | Number of Votes |
| Mystery | 12 |
| Biography | 16 |
| Fantasy | 20 |
| Science Fiction | 8 |

a. Draw tape diagrams with a unit size of 4 to represent the book club's favorite type of book.

b. Use your tape diagrams to draw vertical tape diagrams that represent the data.

Lesson 2:   Rotate tape diagrams vertically.

29

Name _____  Date _____

1.  Adi surveys third graders to find out their favorite fruits.  The results are in the table below.

| Favorite Fruits of Third Graders | |
| --- | --- |
| Fruit | Number of Student Votes |
| Banana | 8 |
| Apple | 16 |
| Strawberry | 12 |
| Peach | 4 |

Draw units of 2 to complete the tape diagrams to show the total votes for each fruit.  The first one has been done for you.

Banana:   | 2 | 2 | 2 | 2 |

Apple:

Strawberry:

Peach:

2.  Explain how you can create vertical tape diagrams to show this data.

Lesson 2:       Rotate tape diagrams vertically.

©2015 Great Minds. eureka-math.org
G3-M6-TE-B6-1.3.1-01.2016

EUREKA
MATH™

3.  Complete the vertical tape diagrams below using the data from Problem 1.

a.

| 2 |
| 2 |
| 2 |
| 2 |

Banana    Apple    Strawberry    Peach

b.

| 4 |
| 4 |

Banana    Apple    Strawberry    Peach

c.  What is a good title for the vertical tape diagrams?

d.  Compare the number of units used in the vertical tape diagrams in Problems 3(a) and 3(b). Why does the number of units change?

e.  Write a multiplication number sentence to show the total number of votes for strawberry in the vertical tape diagram in Problem 3(a).

f.  Write a multiplication number sentence to show the total number of votes for strawberry in the vertical tape diagram in Problem 3(b).

g.  What changes in your multiplication number sentences in Problems 3(e) and (f)? Why?

# Lesson 3

Objective:  Create scaled bar graphs.

## Suggested Lesson Structure

■ Fluency Practice          (12 minutes)
■ Application Problem        (5 minutes)
■ Concept Development        (33 minutes)
■ Student Debrief           (10 minutes)
  **Total Time**            **(60 minutes)**

## Fluency Practice  (12 minutes)

- How Many Units of 6  **3.OA.1**          (3 minutes)
- Sprint:  Multiply or Divide by 6  **3.OA.4**          (9 minutes)

### How Many Units of 6  (3 minutes)

Note:  This activity reviews multiplication and division with units of 6.

Direct students to count forward and backward by sixes to 60, occasionally changing the direction of the count.

    T:   How many units of 6 are in 12?
    S:   2 units of 6.
    T:   Give me the division sentence with the number of sixes as the quotient.
    S:   $12 \div 6 = 2$.

Continue the process with 24, 36, and 48.

### Sprint:  Multiply or Divide by 6  (9 minutes)

Materials:   (S) Multiply or Divide by 6 Sprint

Note:  This Sprint supports multiplication and division using units of 6.

**Lesson 3:**      Create scaled bar graphs.

©2015 Great Minds. eureka-math.org
G3-M6-TE-B6-1.3.1-01.2016

EUREKA
MATH™

## Application Problem (5 minutes)

The vertical tape diagrams show the number of fish in Sal's Pet Store.

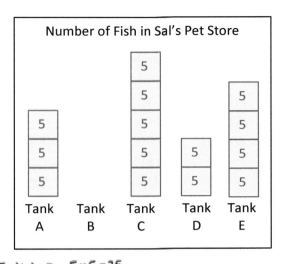

Number of Fish in Sal's Pet Store

a.  Find the total number of fish in Tank C.
    Show your work.

b.  Tank B has a total of 30 fish. Draw the tape diagram for Tank B.

c.  How many more fish are in Tank B than in Tanks A and D combined?

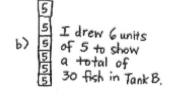

a) 5×5 = 25
    There are 25 fish
    in Tank C.

b) I drew 6 units
   of 5 to show
   a total of
   30 fish in Tank B.

c) Tank A+D: 5×5 = 25
   30 − 25 = 5
   There are 5 more fish
   in Tank B than Tanks
   A and D combined.

Note: This problem reviews reading vertical tape diagrams with a unit size larger than 1. It also anticipates the Concept Development, where students construct a scaled bar graph from the data in this problem.

## Concept Development (33 minutes)

Materials:  (S) Graph A (Template 1) pictured below, Graph B (Template 2) pictured below, colored pencils, straightedge

**Problem 1: Construct a scaled bar graph.**

Template 1 with Student Work

T:  (Pass out Template 1 pictured to the right.) Draw the vertical tape diagrams from the Application Problem on the grid. (Allow students time to work.) Outline the bars with your colored pencil. Erase the unit labels inside the bar, and shade the entire bar with your colored pencil. (Model an example.)

T:  What does each square on the grid represent?

S:  5 fish!

T:  We can show that by creating a scale on our bar graph. (Write 0 where the axes intersect, and then write 5 near the first line on the vertical axis. Point to the next line up on the grid.) Turn and talk to a partner. What number should I write here? How do you know?

S:  Ten because you are counting by fives. → Ten because each square has a value of 5, and 2 fives is 10.

T:  Count by fives to complete the rest of the scale on the graph.

**MP.6**

S:  (Count and write.)

T:  What do the numbers on the scale tell you?

S:  The number of fish!

T:  Label the scale *Number of Fish*. (Model.) What do the labels under each bar tell you?

S:  Which tank the bar is for!

T:  What is a good title for this graph?

S:  *Number of Fish at Sal's Pet Store*.

T:  Write the title *Number of Fish at Sal's Pet Store*. (Model.)

T:  Turn and talk to a partner. How is this **scaled bar graph** similar to the vertical tape diagrams in the Application Problem? How is it different?

S:  They both show the number of fish in Sal's pet store. → The value of the bars and the tape diagrams is the same. → The way we show the value of the bars changed. In the Application Problem, we labeled each unit. In this graph, we made a scale to show the value.

T:  You are right. This scaled bar graph does not have labeled units, but it has a scale we can read to find the values of the bars. (Pass out Template 2, pictured to the right.) Let's create a second bar graph from the data. What do you notice about the labels on this graph?

S:  They are switched! → Yeah, the tank labels are on the side, and the *Number of Fish* label is now at the bottom.

T:  Count by fives to label your scale along the horizontal edge. Then, shade in the correct number of squares for each tank. Will your bars be horizontal or vertical?

S:  Horizontal. (Label and shade.)

T:  Take Graph A and turn it so the paper is horizontal. Compare it with Graph B. What do you notice?

S:  They are the same!

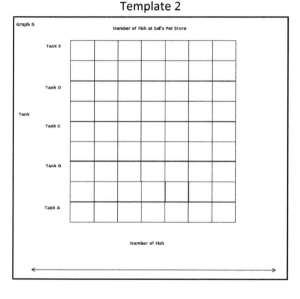

T:  A bar graph can be drawn vertically or horizontally, depending on where you decide to put the labels, but the information stays the same as long as the scales are the same.

T:  Marcy buys 3 fish from Tank C. Write a subtraction sentence to show how many fish are left in Tank C.

S:  (Write 25 − 3 = 22.)

T:  How many fish are left in Tank C?

S:  22 fish!

T:  Discuss with a partner how I can show 22 fish on the bar graph.

S:  (Discuss.)

T:  I am going to erase some of the Tank C bar. Tell me to stop when you think it shows 22 fish. (Erase until students say to stop.) Even though our scale counts by fives, we can show other values for the bars by drawing the bars in between the numbers on the scale.

**Problem 2: Plot data from a bar graph on a number line.**

T:  Let's use Graph B to create a number line to show the same information. There is an empty number line below the graph. Line up a straightedge with each column on the grid to make intervals on the number line that match the scale on the graph. (Model.)

S:  (Draw intervals.)

T:  Should the intervals on the number line be labeled with the number of fish or with the tanks? Discuss with your partner.

S:  The number of fish.

T:  Why? Talk to your partner.

S:  The number of fish because the number line shows the scale.

T:  Label the intervals. (Allow students time to work.) Now, work with a partner to plot and label the number of fish in each tank on the number line.

S:  (Plot and label.)

T:  Talk to a partner. Compare how the information is shown on the bar graph and the number line.

S:  The tick marks on the number line are in the same places as the graph's scale. → The spaces in between the tick marks on the number line are like the unit squares on the bar graph. → On the number line, the tanks are just dots, not whole bars, so the labels look a little different, too.

T:  We can read different information from the 2 representations. Compare the information we can read.

S:  With a bar graph, it is easy to see the order from least to most fish just by looking at the size of the bars. → The number line shows you how much, too, but you know which is the most by looking for the biggest number on the line, not by looking for the biggest bar.

T:  Yes. A bar graph allows us to compare easily. A number line plots the information.

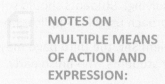

**NOTES ON MULTIPLE MEANS OF ACTION AND EXPRESSION:**

Assist students with perceptual difficulties, low vision, and others with plotting corresponding points on the number line. To make tick marks, show students how to hold and align the straightedge with the scale at the bottom of the graph, *not the bars.* Precise alignment is desired, but comfort, confidence, accurate presentation of data, and a frustration-free experience are more valuable.

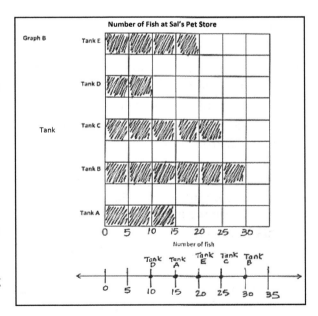

## Problem Set  (10 minutes)

Students should do their personal best to complete the Problem Set within the allotted 10 minutes.  For some classes, it may be appropriate to modify the assignment by specifying which problems they work on first.  Some problems do not specify a method for solving.  Students should solve these problems using the RDW approach used for Application Problems.

For this Problem Set, the third page may be used as an extension for students who finish early.

## Student Debrief  (10 minutes)

**Lesson Objective:**  Create scaled bar graphs.

The Student Debrief is intended to invite reflection and active processing of the total lesson experience.

Invite students to review their solutions for the Problem Set. They should check work by comparing answers with a partner before going over answers as a class.  Look for misconceptions or misunderstandings that can be addressed in the Debrief. Guide students in a conversation to debrief the Problem Set and process the lesson.

Any combination of the questions below may be used to lead the discussion.

- Discuss your simplifying strategy, or a simplifying strategy you could have used, for Problem 1(b).
- Share number sentences for Problem 1(c).
- How did the straightedge help you read the bar graph in Problem 2?
- Share your number line for Problem 4.  How did the scale on the bar graph help you draw the intervals on the number line?  What does each interval on the number line represent?
- Did you use the bar graph or the number line to answer the questions in Problem 5?  Explain your choice.
- Compare vertical tape diagrams to **scaled bar graphs**.  (If necessary, clarify the phrase *scaled bar graph*.)  What is different?  What is the same?

NOTES ON
MULTIPLE MEANS
OF REPRESENTATION:

Students working below grade level and others may benefit from the following scaffolds for reading graphs on the Problem Set:

- Facilitate a guided practice of estimating and accurately determining challenging bar values. Start with smaller numbers and labeled increments, gradually increasing the challenge.

- Draw, or have students draw, a line (in a color other than black) aligning the top of the bar with its corresponding measure on the scale.

- Allow students to record the value inside of the bar—in increments as a tape diagram or as a whole—until they become proficient.

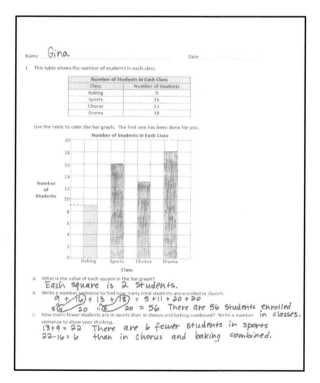

©2015 Great Minds. eureka-math.org
G3-M6-TE-B6-1.3.1-01.2016

- Does the information change when a bar graph is drawn horizontally or vertically with the same scale?  Why or why not?
- What is the purpose of a label on a bar graph?
- How is a bar graph's scale more precise than a picture graph's?
- How does the fluency activity, Group Counting on a Vertical Number Line, relate to reading a bar graph?

## Exit Ticket  (3 minutes)

After the Student Debrief, instruct students to complete the Exit Ticket.  A review of their work will help with assessing students' understanding of the concepts that were presented in today's lesson and planning more effectively for future lessons.  The questions may be read aloud to the students.

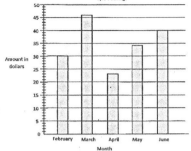

2. This bar graph shows Kyle's savings from February to June. Use a straight edge to help you read the graph.

a. How much money did Kyle save in May?  He saved $34 in May.

b. In which months did Kyle save less than $35?  February, April, and May.

c. How much more did Kyle save in June than April? Write a number sentence to show your thinking.  $40 − $23 = $17

d. The money Kyle saved in _April_ was half the money he saved in _March_.

3. Complete the table below to show the same data given in the bar graph in problem 2.

| Months | February | March | April | May | June |
|---|---|---|---|---|---|
| Amount in dollars saved | $30 | $46 | $23 | $34 | $40 |

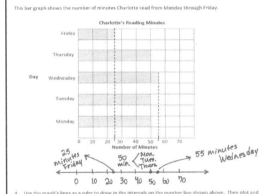

This bar graph shows the number of minutes Charlotte read from Monday through Friday.

4. Use the graph's lines as a ruler to draw in the intervals on the number line shown above.  Then plot and label a point for each day on the number line.

5. Use the graph or number line to answer the following questions.

a. On which days did Charlotte read for the same number of minutes?  How many minutes did Charlotte read on these days?  Charlotte read for 50 minutes on Monday, Tuesday, and Thursday.

b. How many more minutes did Charlotte read on Wednesday than on Friday?  55 − 25 = 30  Charlotte read 30 minutes more on Wednesday than on Friday.

# A

Number Correct: _____

Multiply or Divide by 6

| | | |
|---|---|---|
| 1. | 2 × 6 = | |
| 2. | 3 × 6 = | |
| 3. | 4 × 6 = | |
| 4. | 5 × 6 = | |
| 5. | 1 × 6 = | |
| 6. | 12 ÷ 6 = | |
| 7. | 18 ÷ 6 = | |
| 8. | 30 ÷ 6 = | |
| 9. | 6 ÷ 6 = | |
| 10. | 24 ÷ 6 = | |
| 11. | 6 × 6 = | |
| 12. | 7 × 6 = | |
| 13. | 8 × 6 = | |
| 14. | 9 × 6 = | |
| 15. | 10 × 6 = | |
| 16. | 48 ÷ 6 = | |
| 17. | 42 ÷ 6 = | |
| 18. | 54 ÷ 6 = | |
| 19. | 36 ÷ 6 = | |
| 20. | 60 ÷ 6 = | |
| 21. | ___ × 6 = 30 | |
| 22. | ___ × 6 = 6 | |

| | | |
|---|---|---|
| 23. | ___ × 6 = 60 | |
| 24. | ___ × 6 = 12 | |
| 25. | ___ × 6 = 18 | |
| 26. | 60 ÷ 6 = | |
| 27. | 30 ÷ 6 = | |
| 28. | 6 ÷ 6 = | |
| 29. | 12 ÷ 6 = | |
| 30. | 18 ÷ 6 = | |
| 31. | ___ × 6 = 36 | |
| 32. | ___ × 6 = 42 | |
| 33. | ___ × 6 = 54 | |
| 34. | ___ × 6 = 48 | |
| 35. | 42 ÷ 6 = | |
| 36. | 54 ÷ 6 = | |
| 37. | 36 ÷ 6 = | |
| 38. | 48 ÷ 6 = | |
| 39. | 11 × 6 = | |
| 40. | 66 ÷ 6 = | |
| 41. | 12 × 6 = | |
| 42. | 72 ÷ 6 = | |
| 43. | 14 × 6 = | |
| 44. | 84 ÷ 6 = | |

Lesson 3:     Create scaled bar graphs.

EUREKA MATH™

# B

Number Correct: _____

Improvement: _____

**Multiply or Divide by 6**

| | | |
|---|---|---|
| 1. | 1 × 6 = | |
| 2. | 2 × 6 = | |
| 3. | 3 × 6 = | |
| 4. | 4 × 6 = | |
| 5. | 5 × 6 = | |
| 6. | 18 ÷ 6 = | |
| 7. | 12 ÷ 6 = | |
| 8. | 24 ÷ 6 = | |
| 9. | 6 ÷ 6 = | |
| 10. | 30 ÷ 6 = | |
| 11. | 10 × 6 = | |
| 12. | 6 × 6 = | |
| 13. | 7 × 6 = | |
| 14. | 8 × 6 = | |
| 15. | 9 × 6 = | |
| 16. | 42 ÷ 6 = | |
| 17. | 36 ÷ 6 = | |
| 18. | 48 ÷ 6 = | |
| 19. | 60 ÷ 6 = | |
| 20. | 54 ÷ 6 = | |
| 21. | ___ × 6 = 6 | |
| 22. | ___ × 6 = 30 | |

| | | |
|---|---|---|
| 23. | ___ × 6 = 12 | |
| 24. | ___ × 6 = 60 | |
| 25. | ___ × 6 = 18 | |
| 26. | 12 ÷ 6 = | |
| 27. | 6 ÷ 6 = | |
| 28. | 60 ÷ 6 = | |
| 29. | 30 ÷ 6 = | |
| 30. | 18 ÷ 6 = | |
| 31. | ___ × 6 = 18 | |
| 32. | ___ × 6 = 24 | |
| 33. | ___ × 6 = 54 | |
| 34. | ___ × 6 = 42 | |
| 35. | 48 ÷ 6 = | |
| 36. | 54 ÷ 6 = | |
| 37. | 36 ÷ 6 = | |
| 38. | 42 ÷ 6 = | |
| 39. | 11 × 6 = | |
| 40. | 66 ÷ 6 = | |
| 41. | 12 × 6 = | |
| 42. | 72 ÷ 6 = | |
| 43. | 13 × 6 = | |
| 44. | 78 ÷ 6 = | |

Name _____    Date _____

1.  This table shows the number of students in each class.

| Number of Students in Each Class | |
| --- | --- |
| Class | Number of Students |
| Baking | 9 |
| Sports | 16 |
| Chorus | 13 |
| Drama | 18 |

Use the table to color the bar graph.  The first one has been done for you.

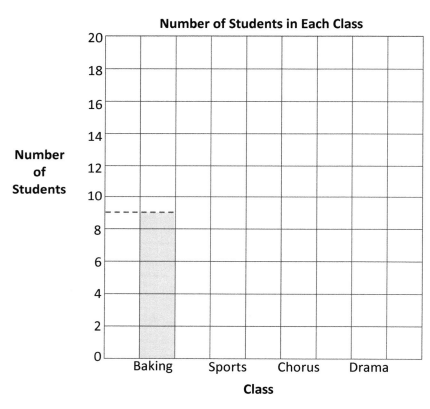

a.  What is the value of each square in the bar graph?

b.  Write a number sentence to find how many total students are enrolled in classes.

c.  How many fewer students are in sports than in chorus and baking combined?  Write a number sentence to show your thinking.

EUREKA
MATH

2.  This bar graph shows Kyle's savings from February to June.  Use a straightedge to help you read the graph.

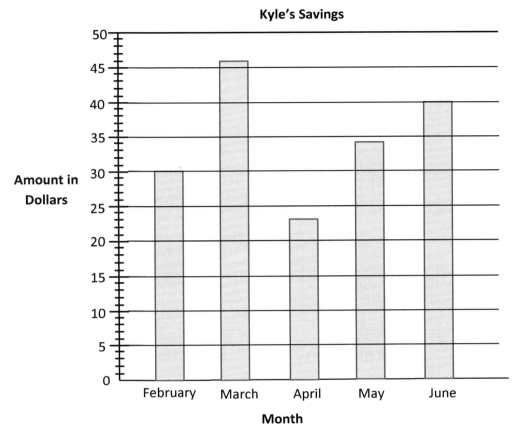

a.  How much money did Kyle save in May?

b.  In which months did Kyle save less than $35?

c.  How much more did Kyle save in June than April?  Write a number sentence to show your thinking.

d.  The money Kyle saved in _____ was half the money he saved in _____.

3.  Complete the table below to show the same data given in the bar graph in Problem 2.

| Months | February | | | | |
|---|---|---|---|---|---|
| Amount Saved in Dollars | | | | | |

This bar graph shows the number of minutes Charlotte read from Monday through Friday.

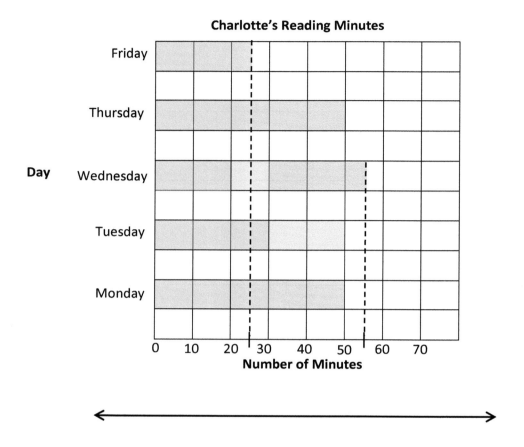

**Charlotte's Reading Minutes**

4. Use the graph's lines as a ruler to draw in the intervals on the number line shown above.  Then plot and label a point for each day on the number line.

5. Use the graph or number line to answer the following questions.

   a. On which days did Charlotte read for the same number of minutes?  How many minutes did Charlotte read on these days?

   b. How many more minutes did Charlotte read on Wednesday than on Friday?

EUREKA
MATH™

Name _____ Date _____

The bar graph below shows the students' favorite ice cream flavors.

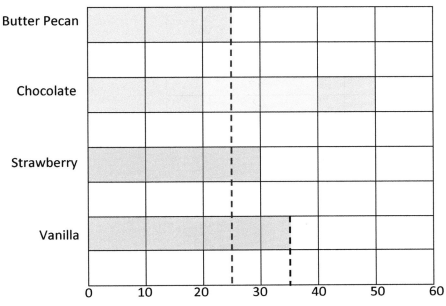

**Favorite Ice Cream Flavors**

**Flavor**: Butter Pecan, Chocolate, Strawberry, Vanilla

**Number of Students**

a. Use the graph's lines as a ruler to draw intervals on the number line shown above. Then plot and label a point for each flavor on the number line.

b. Write a number sentence to show the total number of students who voted for butter pecan, vanilla, and chocolate.

Name _____   Date _____

1.   This table shows the favorite subjects of third graders at Cayuga Elementary.

| Favorite Subjects | |
|---|---|
| Subject | Number of Student Votes |
| Math | 18 |
| ELA | 13 |
| History | 17 |
| Science | ? |

Use the table to color the bar graph.

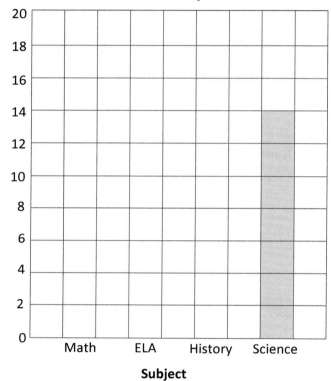

**Favorite Subjects**

a.   How many students voted for science?

b.   How many more students voted for math than for science?  Write a number sentence to show your thinking.

c.   Which gets more votes, math and ELA together or history and science together?  Show your work.

Lesson 3:      Create scaled bar graphs.

EUREKA
MATH™

2. This bar graph shows the number of liters of water Skyler uses this month.

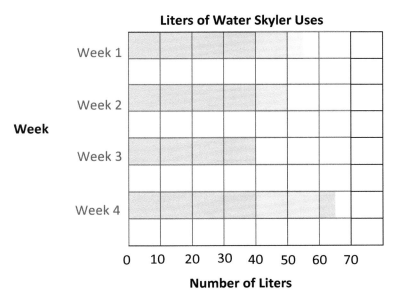

a. During which week does Skyler use the most water? _____
   The least? _____

b. How many more liters does Skyler use in Week 4 than Week 2?

c. Write a number sentence to show how many liters of water Skyler uses during Weeks 2 and 3 combined.

d. How many liters does Skyler use in total?

e. If Skyler uses 60 liters in each of the 4 weeks next month, will she use more or less than she uses this month? Show your work.

3. Complete the table below to show the data displayed in the bar graph in Problem 2.

| Liters of Water Skyler Uses | |
| --- | --- |
| Week | Liters of Water |
| | |
| | |
| | |
| | |

EUREKA
MATH™

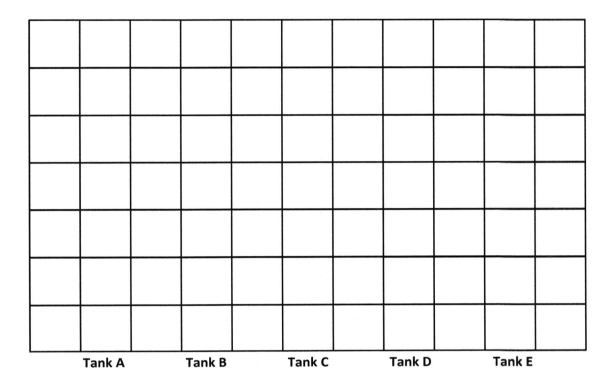

                    Tank A          Tank B          Tank C          Tank D          Tank E

                                        **Tank**

graph A

**Number of Fish at Sal's Pet Store**

Tank E

Tank D

Tank

Tank C

Tank B

Tank A

**Number of Fish**

graph B

# Lesson 4

Objective:  Solve one- and two-step problems involving graphs.

## Suggested Lesson Structure

■ Fluency Practice          (10 minutes)
■ Application Problem        (8 minutes)
□ Concept Development        (32 minutes)
■ Student Debrief           (10 minutes)

    **Total Time**          **(60 minutes)**

## Fluency Practice  (10 minutes)

- Read Line Plots  **2.MD.9**      (5 minutes)
- Read Bar Graphs  **3.MD.3**      (5 minutes)

## Read Line Plots  (5 minutes)

Materials:   (T) Line plot (Fluency Template 1) pictured to the right
             (S) Personal white board

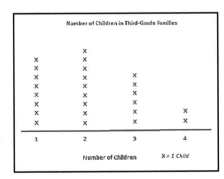

Note:  This activity reviews Grade 2 concepts about line plots in preparation for Topic B.

  T:  (Project the line plot.)  This line plot shows how many children are in the families of students in a third-grade class.  How many students only have one child in their family?  Let's count to find the answer.  (Point to the X's as students count.)

  S:  1, 2, 3, 4, 5, 6, 7, 8.

Continue the process for 2 children, 3 children, and 4 children.

  T:  Most students have how many children in their family?

  S:  2 children.

  T:  On your personal white boards, write a number sentence to show how many more third graders have 2 children in their family than 3 children.

  S:  (Write 9 – 6 = 3.)

Continue the process to find how many fewer third graders have 4 children in their family than 2 children and how many more third graders have 1 child in their family than 3 children.

**NOTES ON
MULTIPLE MEANS
OF REPRESENTATION:**

Scaffold for English language learners and others how to solve for *how many more*.  Ask, "How many third graders have 2 children in their family?  How many have 3 children?  Which is more, 6 or 9?  How many more? (Count up from 6 to 9)."

T: On your board, write a number sentence to show how many third graders have 3 or 4 children in their family.

S: (Write 6 + 2 = 8.)

Continue the process to find how many third graders have 1 or 2 children in their family and how many third graders have a sibling.

## Read Bar Graphs  (5 minutes)

Materials: (T) Bar graph (Fluency Template 2) pictured to the right  (S) Personal white board

Notes: This activity reviews Lesson 3.

T: (Project the bar graph Template.)  This bar graph shows how many minutes 4 children spent practicing piano.

T: Did Ryan practice for more or less than 30 minutes?

S: More.

T: Did he practice for more or less than 40 minutes?

S: Less.

T: What fraction of the time between 30 and 40 minutes did Ryan practice piano?

S: 1 half of the time.

T: What is halfway between 30 minutes and 40 minutes?

S: 35 minutes.

T: The dotted line is there to help you read 35 since 35 is between two numbers on the graph.  How long did Kari spend practicing piano?

S: 40 minutes.

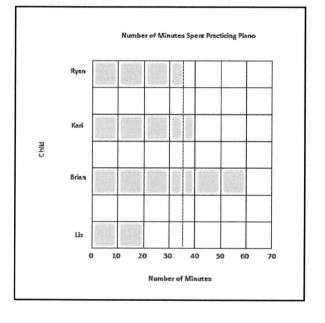

Continue the process for Brian and Liz.

T: Who practiced the longest?

S: Brian.

T: Who practiced the least amount of time?

S: Liz.

T: On your personal white board, write a number sentence to show how much longer Brian practiced than Kari.

S: (Write 60 − 40 = 20.)

Continue the process to find how many fewer minutes Ryan practiced than Brian.

T: On your board, write a number sentence to show how many total minutes Kari and Liz spent practicing piano.

S: (Write 40 + 20 = 60.)

Continue the process to find how many total minutes Ryan and Brian spent practicing piano and how many total minutes all the children practiced.

Lesson 4:  Solve one- and two-step problems involving graphs.

EUREKA MATH™

## Application Problem  (8 minutes)

The following chart shows the number of times an insect's wings vibrate each second.  Use the following clues to complete the unknowns in the chart.

a.  The beetle's number of wing vibrations is the same as the difference between the fly's and honeybee's.

b.  The mosquito's number of wing vibrations is the same as 50 less than the beetle's and fly's combined.

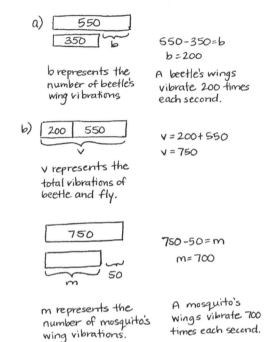

a)  550
    350      $550 - 350 = b$
             $b = 200$

b represents the number of beetle's wing vibrations

A beetle's wings vibrate 200 times each second.

b)  200  550    $v = 200 + 550$
                $v = 750$

v represents the total vibrations of beetle and fly.

750          $750 - 50 = m$
             $m = 700$
    m    50

m represents the number of mosquito's wing vibrations.

A mosquito's wings vibrate 700 times each second.

| Wing Vibrations of Insects | |
| --- | --- |
| Insect | Number of Wing Vibrations Each Second |
| Honeybee | 350 |
| Beetle | $b$ |
| Fly | 550 |
| Mosquito | $m$ |

Note:  The data from the chart is used in the upcoming Concept Development, where students first create a bar graph and then answer one- and two-step questions from the graph.

### Graph Template

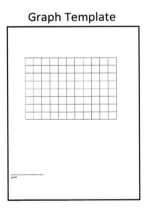

## Concept Development  (32 minutes)

Materials:   (S) Graph (Template) pictured to the right, personal white board

T:  (Pass out the graph Template.)  Let's create a bar graph from the data in the Application Problem.  We need to choose a scale that works for the data the graph represents.  Talk to a partner:  What scale would be best for this data?  Why?

S:  We could count by fives or tens.  → The numbers are big, so that would be a lot of tick marks to draw.  → We could do it by hundreds since the numbers all end in zero.

T:  In this case, using hundreds is a strong choice since the numbers are between 200 and 700.  Decide if you will show the scale for your graph vertically or horizontally.  Then, label it starting at zero.

S:  (Label.)

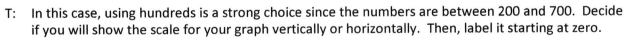

Lesson 4:  Solve one- and two-step problems involving graphs.

©2015 Great Minds. eureka-math.org
G3-M6-TE-B6-1.3.1-01.2016

T:  The number of wing vibrations for the honeybee is 350 each second.  Discuss the bar you will make for the honeybee with your partner.  How many units will you shade in?

S:  Maybe 4 units.  We can round up.  → But to show the exact number, we just need to shade in 3 and one-half units.

T:  Many of you noticed that you need to shade a half unit to show this data precisely.  Do you need to do the same for other insects?

S:  We also have to do this for the fly since it is 550.

T:  Go ahead and shade your bars.

S:  (Shade bars.)

T:  On your personal white board, write a number sentence to find the total number of vibrations 2 beetles and 1 honeybee can produce each second.

S:  (350 + 200 + 200 = 750.)

T:  Use a tape diagram to compare how many more vibrations a fly and honeybee combined produce than a mosquito.

S:  (Work should resemble the sample below.)

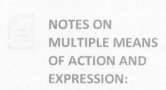

NOTES ON MULTIPLE MEANS OF ACTION AND EXPRESSION:

Scaffold partner talk with sentence frames such as the ones listed below.

- I notice _____.
- The _____'s wings are faster than the _____'s.
- When I compare the _____ and _____, I see that …
- I did not know that…
- This data is interesting because…

MP.3

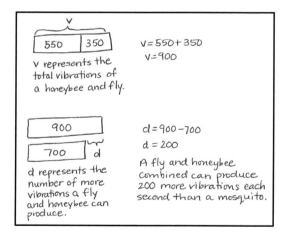

T:  Work with your partner to think of another question that can be solved using the data on this graph.  Solve your question, and then trade questions with the pair of students next to you.  Solve the new question, and check your work with their work.

## Problem Set  (10 minutes)

Students should do their personal best to complete the Problem Set within the allotted 10 minutes.  For some classes, it may be appropriate to modify the assignment by specifying which problems they work on first.  Some problems do not specify a method for solving.  Students should solve these problems using the RDW approach used for Application Problems.

> **NOTES ON THE PROBLEM SET:**
>
> Problem 1(a) on the Problem Set may be the first time your students create a bar graph without the scaffold of a grid.  Bring this to students' attention, and quickly review how the bars should be created.

## Student Debrief  (10 minutes)

**Lesson Objective:**  Solve one- and two-step problems involving graphs.

The Student Debrief is intended to invite reflection and active processing of the total lesson experience.

Invite students to review their solutions for the Problem Set.  They should check work by comparing answers with a partner before going over answers as a class.  Look for misconceptions or misunderstandings that can be addressed in the Debrief.  Guide students in a conversation to debrief the Problem Set and process the lesson.

Any combination of the questions below may be used to lead the discussion.

- Invite students who used different scales for Problem 1 to share their work.
- How did you solve Problem 1(c)?  What did you do first?
- What is the value of each interval in the bar graph in Problem 2?  How do you know?
- How did you solve Problem 2(a)?
- Explain to your partner what you needed to do before answering Problem 2(b).
- Compare the chart from the Application Problem with the bar graph you made of that same data.  How is each representation a useful tool?  When might you choose to use each representation?
- How did the fluency activity, Read Bar Graphs, help you get ready for today's lesson?

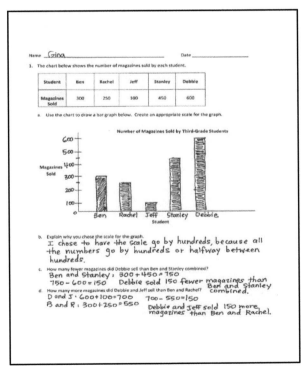

## Exit Ticket  (3 minutes)

After the Student Debrief, instruct students to complete the Exit Ticket.  A review of their work will help with assessing students' understanding of the concepts that were presented in today's lesson and planning more effectively for future lessons.  The questions may be read aloud to the students.

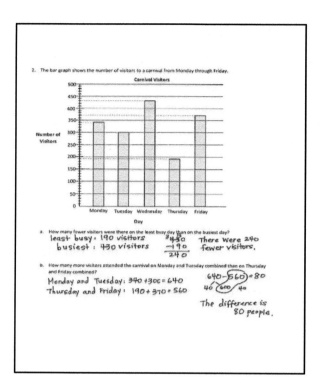

EUREKA
MATH™

Name _____ Date _____

1. The chart below shows the number of magazines sold by each student.

| Student | Ben | Rachel | Jeff | Stanley | Debbie |
|---------|-----|--------|------|---------|--------|
| Magazines Sold | 300 | 250 | 100 | 450 | 600 |

a. Use the chart to draw a bar graph below. Create an appropriate scale for the graph.

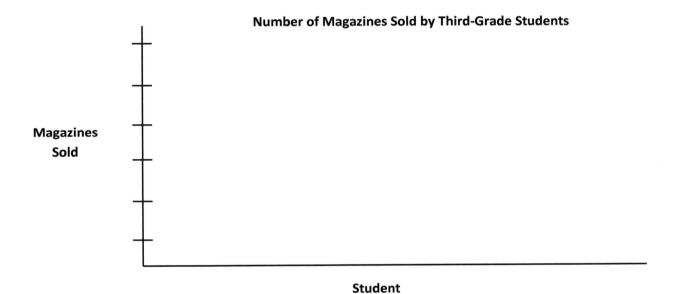

b. Explain why you chose the scale for the graph.

c. How many fewer magazines did Debbie sell than Ben and Stanley combined?

d. How many more magazines did Debbie and Jeff sell than Ben and Rachel?

EUREKA MATH™

Lesson 4: Solve one- and two-step problems involving graphs.

55

©2015 Great Minds. eureka-math.org
G3-M6-TE-B6-1.3.1-01.2016

2. The bar graph shows the number of visitors to a carnival from Monday through Friday.

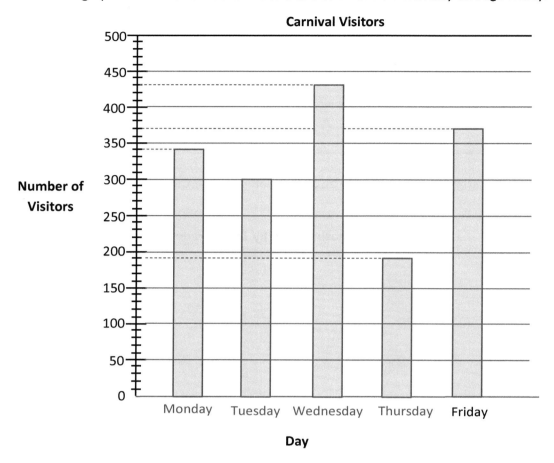

**Carnival Visitors**

a. How many fewer visitors were there on the least busy day than on the busiest day?

b. How many more visitors attended the carnival on Monday and Tuesday combined than on Thursday and Friday combined?

Lesson 4: Solve one- and two-step problems involving graphs.

©2015 Great Minds. eureka-math.org
G3-M6-TE-B6-1.3.1-01.2016

**EUREKA MATH**™

Name _____ Date _____

The graph below shows the number of library books checked out in five days.

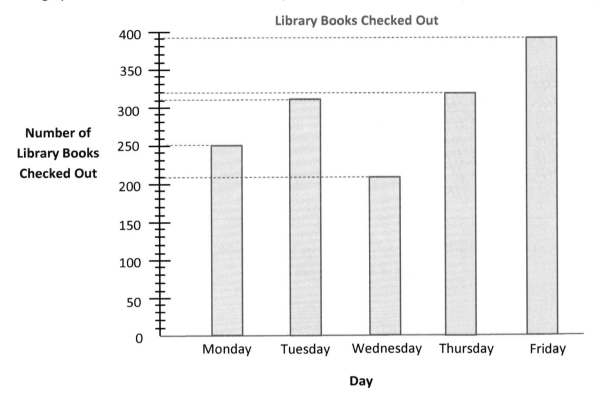

**Library Books Checked Out**

Number of Library Books Checked Out

Day

c.  How many books in total were checked out on Wednesday and Thursday?

d.  How many more books were checked out on Thursday and Friday than on Monday and Tuesday?

Name _____     Date _____

1.  Maria counts the coins in her piggy bank and records the results in the tally chart below.  Use the tally marks to find the total number of each coin.

| Coins in Maria's Piggy Bank | | |
| --- | --- | --- |
| **Coin** | **Tally** | **Number of Coins** |
| Penny | ℍℍ ℍℍ ℍℍ ℍℍ ℍℍ ℍℍ ℍℍ ℍℍ ℍℍ ℍℍ ℍℍ ℍℍ ℍℍ /// | |
| Nickel | ℍℍ ℍℍ ℍℍ ℍℍ ℍℍ ℍℍ ℍℍ ℍℍ ℍℍ ℍℍ ℍℍ ℍℍ // | |
| Dime | ℍℍ ℍℍ ℍℍ ℍℍ ℍℍ ℍℍ ℍℍ ℍℍ ℍℍ ℍℍ ℍℍ // | |
| Quarter | ℍℍ ℍℍ ℍℍ ℍℍ //// | |

a.  Use the tally chart to complete the bar graph below.  The scale is given.

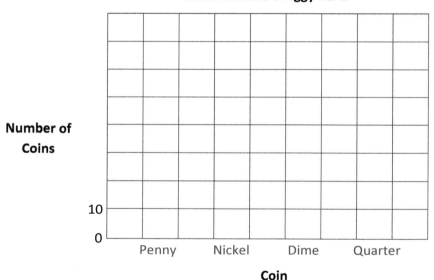

**Coins in Maria's Piggy Bank**

Number of Coins

10

0

Penny    Nickel    Dime    Quarter

**Coin**

b.  How many more pennies are there than dimes?

c.  Maria donates 10 of each type of coin to charity.  How many total coins does she have left?  Show your work.

EUREKA
MATH

2.  Ms. Hollmann's class goes on a field trip to the planetarium with Mr. Fiore's class.  The number of students in each class is shown in the picture graphs below.

| Students in Ms. Hollmann's Class |
|---|
| **Boys** ☐☐☐☐☐☐☐ |
| **Girls** ☐☐☐☐☐☐☐☐ |
| ☐ = 2 students |

| Students in Mr. Fiore's Class |
|---|
| **Boys** ☐☐☐☐☐☐☐ |
| **Girls** ☐☐☐☐☐☐☐☐ |
| ☐ = 2 students |

a.  How many fewer boys are on the trip than girls?

b.  It costs $2 for each student to attend the field trip.  How much money does it cost for all students to attend?

c.  The cafeteria in the planetarium has 9 tables with 8 seats at each table.  Counting students and teachers, how many empty seats should there be when the 2 classes eat lunch?

EUREKA
MATH™

Lesson 4:    Solve one- and two-step problems involving graphs.

©2015 Great Minds. eureka-math.org
G3-M6-TE-B6-1.3.1-01.2016

59

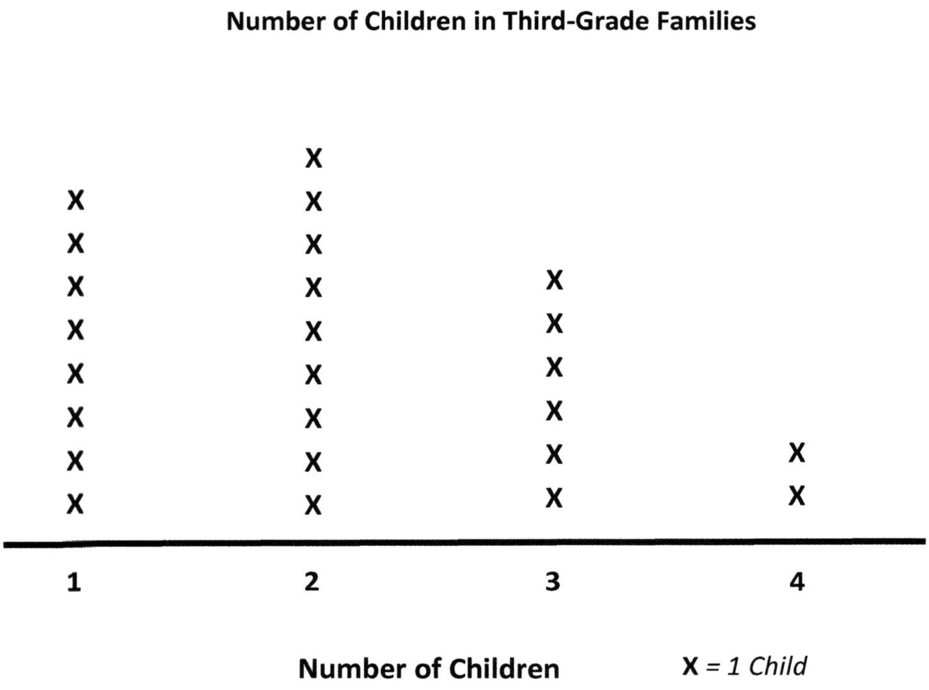

**Number of Children in Third-Grade Families**

```
                 X
      X          X
      X          X
      X          X          X
      X          X          X
      X          X          X
      X          X          X
      X          X          X          X
      X          X          X          X
 ─────────────────────────────────────────────
      1          2          3          4
```

**Number of Children**     **X** = *1 Child*

─────────────────────────

line plot

**Lesson 4:**   Solve one- and two-step problems involving graphs.

EUREKA
MATH™

**Number of Minutes Spent Practicing Piano**

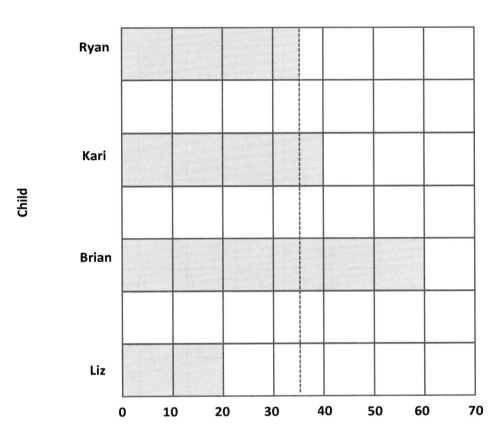

Child

Ryan

Kari

Brian

Liz

0    10    20    30    40    50    60    70

**Number of Minutes**

_____

bar graph

graph

**Lesson 4:** Solve one- and two-step problems involving graphs.

EUREKA
MATH™

**3**
GRADE

# Mathematics Curriculum

## Topic B

# Generate and Analyze Measurement Data

### 3.MD.4

| Focus Standard: | 3.MD.4 | Generate measurement data by measuring lengths using rulers marked with halves and fourths of an inch. Show the data by making a line plot, where the horizontal scale is marked off in appropriate units—whole numbers, halves, or quarters. |
|---|---|---|
| Instructional Days: | 5 | |
| Coherence -Links from: | G2–M7 | Problem Solving with Length, Money, and Data |
| | G3–M5 | Fractions as Numbers on the Number Line |
| -Links to: | G4–M2 | Unit Conversions and Problem Solving with Metric Measurement |

In Lesson 5, students use the method of partitioning a whole into equally spaced increments using the number line as a measurement tool (Module 5, Lesson 30) to partition a six-inch strip into 6 equal increments. They repeat the process by partitioning the same strip into 12 equal increments and determine that it shows half-inch intervals. Finally, students partition the strip into 24 equal increments to determine that they have created quarter-inch intervals. The three measurements on the paper strip respectively measure in whole-inch, half-inch, and quarter-inch measurements.

Students use their paper strip as a ruler to measure pre-cut straws that are less than six inches long. As they measure, they make predictions about which of their measurements gives the most accurate data, eventually concluding that it is typically the quarter-inch measurement.

Lesson 6 reintroduces the line plot as a tool for displaying measurement data. While students are familiar with line plots from Grade 2, using fractional values on the line plot is a new concept in this lesson. To prepare students for creating their own line plots in Lessons 7 and 8, Lesson 6 builds foundational experience with representations given in fractional intervals. Students understand the conventions of line plots with fractions and learn to interpret data from them.

In Lessons 7 and 8, students apply the conventions of constructing line plots with fractions to display measurement data. They learn how to represent data when the data set has values of mixed units (i.e., double-digit whole numbers and a fraction). The process of representing their data on line plots naturally evokes student observations about the distribution of the data and leads to solving comparative problems.

In Lesson 9, students analyze both categorical and measurement data to solve problems. Students also explore the functions of different representations—graphs, charts, and line plots—and discuss the appropriateness of each type of representation for different types of data.

This is a perfect opportunity to take advantage of measuring for science-related purposes. For example, if students are germinating and growing bean plants, they may start by measuring the bean seed and then take regular measurements of the plant as it grows. Students might also collect objects from the playground, such as leaves from the same tree or blades of grass. They could talk about why someone might want to measure these objects (e.g., to analyze the health of the tree).

| A Teaching Sequence Toward Mastery to Generate and Analyze Measurement Data |
|---|

**Objective 1:** Create ruler with 1-inch, $\frac{1}{2}$-inch, and $\frac{1}{4}$-inch intervals, and generate measurement data.
(Lesson 5)

**Objective 2:** Interpret measurement data from various line plots.
(Lesson 6)

**Objective 3:** Represent measurement data with line plots.
(Lessons 7–8)

**Objective 4:** Analyze data to problem solve.
(Lesson 9)

# Lesson 5

Objective:  Create ruler with 1-inch, $\frac{1}{2}$-inch, and $\frac{1}{4}$-inch intervals, and generate measurement data.

## Suggested Lesson Structure

■ Fluency Practice          (10 minutes)
▢ Concept Development    (40 minutes)
■ Student Debrief           (10 minutes)
   **Total Time**              **(60 minutes)**

## Fluency Practice  (10 minutes)

- Group Counting  **3.OA.1**        (6 minutes)
- Factors of 12  **3.MD.4**          (4 minutes)

### Group Counting  (6 minutes)

Materials:   (S) Personal white board

Note:  This group counting activity reviews units of 6 and the relationship between multiplication and division.

T:   Count by sixes to 60.  (Write on the board as students count.)
S:   6, 12, 18, 24, 30, 36, 42, 48, 54, 60.

| 6 | 12 | 18 | 24 | 30 | 36 | 42 | 48 | 54 | 60 |
|---|----|----|----|----|----|----|----|----|----|
| 1 six | 2 sixes | 3 sixes | 4 sixes | 5 sixes | 6 sixes | 7 sixes | 8 sixes | 9 sixes | 10 sixes |
| 6 ÷ 6 | 12 ÷ 6 | 18 ÷ 6 | 24 ÷ 6 | 30 ÷ 6 | 36 ÷ 6 | 42 ÷ 6 | 48 ÷ 6 | 54 ÷ 6 | 60 ÷ 6 |

T:   (Beneath 6, write 1 six.  Point to the 12.)  12 is the same as how many sixes?
S:   2 sixes.
T:   (Write 2 sixes beneath 12.  Point to the 18.)  18 is the same as how many sixes?
S:   3 sixes.
T:   (Write 3 sixes beneath 18.  Point to 1 six.)  Let's count units of 6.  (Write as students count.)
S:   1 six, 2 sixes, 3 sixes, 4 sixes, 5 sixes, 6 sixes, 7 sixes, 8 sixes, 9 sixes, 10 sixes.
T:   (Point to 60.)  How many sixes are in 60?
S:   10 sixes.

T:   (Beneath 10 sixes, write 60 ÷ 6 = ___.)  What is 60 ÷ 6?

S:   10.

T:   (Write 60 ÷ 6 = 10.  Beneath 1 six, write 6 ÷ 6 = ___.)  On your personal white board, write the number sentence.

S:   (6 ÷ 6 = 1.)

Repeat the process for the rest of the chart.

## Factors of 12  (4 minutes)

Note:  This activity prepares students for today's lesson.

T:   (Write 12 × ___ = 12.)  Say the number sentence, completing the unknown factor.

S:   12 × 1 = 12.

Continue with the following possible sequence:  1 × ___ = 12, 6 × ___ = 12, 4 × ___ = 12, 2 × ___ = 12, and 3 × ___ = 12.

T:   After I say a factor, you say the factor you need to multiply it by to get 12.  The first factor is 1.

S:   12.

T:   6?

S:   2.

T:   4?

S:   3.

T:   12?

S:   1.

T:   3?

S:   4.

## Concept Development  (40 minutes)

Materials:   (S) 1" × 6" strip of yellow construction paper, colored pencils or markers (black, red, and blue), ruler, lined paper (Template), 1 straw pre-cut (vary 1", $\frac{1}{2}$", and $\frac{1}{4}$" lengths among students), Problem Set

**Problem 1: Partition and measure a paper strip into a ruler with whole-inch, half-inch, and quarter-inch measurements.**

T:   (Give each student one copy of the lined paper Template.)  Turn your paper so the margin is horizontal.  Draw a number line on top of the margin.  Mark 0 on the point where I did.  (Model.)

T:   Use your black marker to plot a point at every 4 spaces.  Use the paper's vertical lines to measure the 4 spaces.  Then, label the number line from 0 to 6, making sure there are 4 spaces for each part.  Tell your partner how you know each part is equal.

S:   (Discuss.)

**66**          **Lesson 5:**          Create ruler with 1-inch, ½-inch, and ¼-inch intervals, and generate measurement data.

©2015 Great Minds. eureka-math.org
G3-M6-TE-B6-1.3.1-01.2016

T:  Use a ruler to trace the vertical lines up from your number line to the top of the paper at each point. (Pass out 1 yellow strip to each student.) Lay the yellow strip so that the left end touches the 0 endpoint on the original number line and the right end touches the vertical line that you traced at the number 6 (as shown below).

Creating the Number Line          Measuring Inches

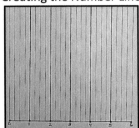    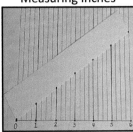

T:  Where the lines touch your strip, plot points on your strip. Extend the points to make them tick marks. Then, turn your strip, and number below each tick mark from 0–6. (After labeling and turning the strip back to its original position, the numbers on the strip will be upside down and ordered opposite from those on the number line. This is shown above.)

T:  Use your ruler to verify that the intervals on your strip are equal. Measure the full length of the yellow strip in inches. Measure the equal parts.

T:  What measurement does each mark represent?

S:  1 inch.

T:  We now know every 4 spaces marks 1 inch on our strip. Let's repeat the process, but this time we will mark a point on our number line (lined paper) at every 2 spaces. What measurement will each mark represent? Talk to a partner.

S:  Two spaces is half. → So, that must mean we will mark half inches!

Repeat the process:

▪  Plot points at every 2 spaces with a red marker to mark half inches. If a point is already marked with a whole inch, plot the new, red point above the black point. Then, plot and label every half inch between the whole inches on the strip.

▪  Plot points at every single interval with a blue marker to mark the quarter inches. If a point is already marked with a whole or half inch, plot the new, blue point above the black or red point. Then, plot every quarter inch between the half inches on the strip. Do not have students label every quarter inch on the strip since the spaces are too small.

**NOTES ON MULTIPLE MEANS OF ACTION AND EXPRESSION:**

Scaffold student partitioning and measuring of the paper strip with the following options:

▪  Instruct step-by-step how to align the zero endpoint of the ruler with the corner of the strip.

▪  Decrease the number of steps by pre-numbering the number line or pre-marking inches.

▪  Use color. Highlight every fourth line of the grid, or lightly shade every other 4 lines.

▪  Make the inch lines tactile with glue to help students with low vision or perceptual difficulties. Because the surface of the grid will be bumpy, have students label numbers once the strip is off the lined paper.

Measuring Half Inches

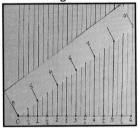

Measuring Quarter Inches

Place the paper strip under a ruler to verify the accuracy of the paper strip's measurements. Encourage students to recognize that their paper strips are, in fact, rulers as well.

T:   Into what three units of measurement did we partition our paper strips, or rulers?

S:   Whole inches, half inches, and quarter inches.

T:   Point to 2 inches on your paper ruler.

S:   (Point.)

T:   Show your partner 1 half inch less than 2 inches on your paper ruler.

S:   (Show.)

T:   What is 1 half inch less than 2 inches?

S:   $1\frac{1}{2}$ inches.

T:   Show $3\frac{1}{4}$ inches.

S:   (Show.)

T:   Show your partner 1 and a quarter inch more than $3\frac{1}{4}$ inches.

S:   (Show.)

T:   What is 1 and a quarter inch more than $3\frac{1}{4}$ inches?

S:   $4\frac{1}{2}$ inches.

Continue the process as needed with $\frac{1}{2}$ inch less than 4 inches, $\frac{1}{4}$ inch more than $1\frac{1}{4}$ inches, $\frac{1}{4}$ inch less than 2 inches, $\frac{3}{4}$ inch more than 3 inches, and $\frac{3}{4}$ inch less than 3 inches.

T:   How many half inches are in 1 inch?

S:   2 half inches.

T:   How many quarter inches are in 1 inch?

S:   4 quarter inches.

T:   How many quarter inches are in 1 half inch?

S:   2 quarter inches.

T:   How many quarter inches are in 3 inches?

S:   12 quarter inches.

**Problem 2: Generate measurement data.**

Pass out the Problem Set and 1 pre-cut straw to each student.

T:   On Problem 1 of your Problem Set, use your paper ruler to measure your straw to the nearest inch, half inch, and quarter inch. What do you do if your measurement is not exact?

S:   We have to estimate.

**68**          **Lesson 5:**          Create ruler with 1-inch, ½-inch, and ¼-inch intervals, and generate
                                          measurement data.

©2015 Great Minds. eureka-math.org
G3-M6-TE-B6-1.3.1-01.2016

EUREKA
MATH™

T:   When you estimate, ask yourself, "Is it more than halfway or less than halfway?"  After measuring the straw you have, measure six of your classmates' straws, and record their measurements in the chart on your Problem Set.

Note:  Students should save their rulers.  They are also used in Lessons 6–7.

## Problem Set  (10 minutes)

Students should do their personal best to complete the remainder of the Problem Set within the allotted 10 minutes. For some classes, it may be appropriate to modify the assignment by specifying which problems they work on first. Some problems do not specify a method for solving.  Students should solve these problems using the RDW approach used for Application Problems.

## Student Debrief  (10 minutes)

**Lesson Objective:**  Create ruler with 1-inch, $\frac{1}{2}$-inch, and $\frac{1}{4}$-inch intervals, and generate measurement data.

The Student Debrief is intended to invite reflection and active processing of the total lesson experience.

Invite students to review their solutions for the Problem Set.  They should check work by comparing answers with a partner before going over answers as a class.  Look for misconceptions or misunderstandings that can be addressed in the Debrief.  Guide students in a conversation to debrief the Problem Set and process the lesson.

Any combination of the questions below may be used to lead the discussion.

- Look at your data for Problem 1.  Did you notice a pattern?
- Share your answer for Problem 1(c).
- Have students share their thinking for Problem 2(c).  If time permits, have a few students measure an object larger than 6 inches with their paper ruler using the method they describe.
- Share your answer to Problem 3.  What number sentence could you use to find the answer?

> **NOTES ON MULTIPLE MEANS OF ACTION AND EXPRESSION:**
>
> Support English language learners as they write their responses on the Problem Set.  Allow students to discuss their thoughts in their language of choice before writing.  Provide sentence starters and a word bank.
>
> Sentence starters may include the following:
>
> - One half inch is the same as_____.
> - It's best to use the quarter-inch ruler to measure because_____.
>
> Possible words for the word bank may include the following:
>
> *exact    estimate    accurate*
>
> *precise   measure*

Name  Gina _____              Date _____

1.  Use the ruler you made to measure different classmates' straws to the nearest inch, $\frac{1}{2}$ inch, and $\frac{1}{4}$ inch. Record the measurements in the chart below.  Draw a star next to measurements that are exact.

| Straw Owner | Measured to the nearest inch | Measured to the nearest $\frac{1}{2}$ inch | Measured to the nearest $\frac{1}{4}$ inch |
|---|---|---|---|
| My straw | 3 | $2\frac{1}{2}$ ★ | $2\frac{1}{4}$ or $2\frac{1}{2}$ ★ |
| Catherine | 4 ★ | 4 ★ | 4 ★ |
| Doug | 2 ★ | 2 ★ | 2 ★ |
| Eva | 4 | $4\frac{1}{2}$ | $4\frac{1}{2}$ |
| Aaron | 3 | $3\frac{1}{2}$ | $3\frac{3}{4}$ |
| Karen | 1 ★ | 1 ★ | 1 ★ |
| Philip | 6 | $5\frac{1}{2}$ | $5\frac{3}{4}$ |

a.  __Karen__'s straw is the shortest straw I measured. It measures __1__ inch(es).

b.  __Philip__'s straw is the longest straw I measured. It measures $5\frac{3}{4}$ inches.

c.  Choose the straw from your chart that was most accurately measured with the $\frac{1}{4}$ inch intervals on your ruler. How do you know the $\frac{1}{4}$ inch intervals are the most accurate for measuring this straw?
Eva's straw was most accurately measured with $\frac{1}{4}$ inch intervals. Measuring to the nearest inch and half inch only gave close estimates, but the quarter inch gave the exact measurement.

Lesson 5:     Create ruler with 1-inch, ½-inch, and ¼-inch intervals, and generate measurement data.

69

©2015 Great Minds. eureka-math.org
G3-M6-TE-B6-1.3.1-01.2016

- How did using the lined paper help you partition your paper strip accurately?

- Each paper strip measured 6 inches, so our measurements were easy to mark. What if the strips were 8 inches instead? How would you partition the number line?

## Exit Ticket  (3 minutes)

After the Student Debrief, instruct students to complete the Exit Ticket. A review of their work will help with assessing students' understanding of the concepts that were presented in today's lesson and planning more effectively for future lessons. The questions may be read aloud to the students.

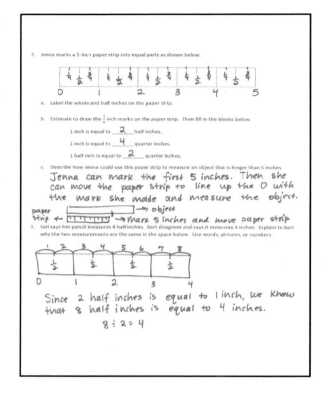

EUREKA
MATH™

Name _____ Date _____

1. Use the ruler you made to measure different classmates' straws to the nearest inch, $\frac{1}{2}$ inch, and $\frac{1}{4}$ inch. Record the measurements in the chart below. Draw a star next to measurements that are exact.

| Straw Owner | Measured to the nearest inch | Measured to the nearest $\frac{1}{2}$ inch | Measured to the nearest $\frac{1}{4}$ inch |
|---|---|---|---|
| My straw | | | |
| | | | |
| | | | |
| | | | |
| | | | |
| | | | |
| | | | |

a. _____'s straw is the shortest straw I measured. It measures _____ inch(es).

b. _____'s straw is the longest straw I measured. It measures _____ inches.

c. Choose the straw from your chart that was most accurately measured with the $\frac{1}{4}$-inch intervals on your ruler. How do you know the $\frac{1}{4}$-inch intervals are the most accurate for measuring this straw?

EUREKA MATH™

Lesson 5: Create ruler with 1-inch, ½-inch, and ¼-inch intervals, and generate measurement data.

©2015 Great Minds. eureka-math.org
G3-M6-TE-B6-1.3.1-01.2016

71

2. Jenna marks a 5-inch paper strip into equal parts as shown below.

   a. Label the whole and half inches on the paper strip.

   b. Estimate to draw the $\frac{1}{4}$-inch marks on the paper strip. Then, fill in the blanks below.

   1 inch is equal to _____ half inches.

   1 inch is equal to _____ quarter inches.

   1 half inch is equal to _____ quarter inches.

   c. Describe how Jenna could use this paper strip to measure an object that is longer than 5 inches.

3. Sari says her pencil measures 8 half inches. Bart disagrees and says it measures 4 inches. Explain to Bart why the two measurements are the same in the space below. Use words, pictures, or numbers.

**Lesson 5:** Create ruler with 1-inch, ½-inch, and ¼-inch intervals, and generate measurement data.

©2015 Great Minds. eureka-math.org
G3-M6-TE-B6-1.3.1-01.2016

Name _____  Date _____

Davon marks a 4-inch paper strip into equal parts as shown below.

a.  Label the whole and quarter inches on the paper strip.

b.  Davon tells his teacher that his paper strip measures 4 inches.  Sandra says it measures 16 quarter inches.  Explain how the two measurements are the same.  Use words, pictures, or numbers.

**Lesson 5:**  Create ruler with 1-inch, ½-inch, and ¼-inch intervals, and generate measurement data.

©2015 Great Minds. eureka-math.org
G3-M6-TE-B6-1.3.1-01.2016

73

Name _____     Date _____

1.  Travis measured 5 different-colored pencils to the nearest inch, $\frac{1}{2}$ inch, and $\frac{1}{4}$ inch.  He records the measurements in the chart below.  He draws a star next to measurements that are exact.

| Colored Pencil | Measured to the nearest inch | Measured to the nearest $\frac{1}{2}$ inch | Measured to the nearest $\frac{1}{4}$ inch |
|---|---|---|---|
| Red | 7 | $6\frac{1}{2}$ | $6\frac{3}{4}$ |
| Blue | 5 | 5 | $5\frac{1}{4}$ |
| Yellow | 6 | $5\frac{1}{2}$ ☆ | $5\frac{1}{2}$ ☆ |
| Purple | 5 | $4\frac{1}{2}$ | $4\frac{3}{4}$ |
| Green | 2 | 3 | $1\frac{3}{4}$ |

a.  Which colored pencil is the longest? _____

    It measures _____ inches.

b.  Look carefully at Travis's data.  Which colored pencil most likely needs to be measured again?  Explain how you know.

Create ruler with 1-inch, ½-inch, and ¼-inch intervals, and generate measurement data.

**EUREKA
MATH**

2. Evelyn marks a 4-inch paper strip into equal parts as shown below.

a. Label the whole and half inches on the paper strip.

b. Estimate to draw the $\frac{1}{4}$-inch marks on the paper strip. Then, fill in the blanks below.

1 inch is equal to _____ half inches.

1 inch is equal to _____ quarter inches

1 half inch is equal to _____ quarter inches.

2 quarter inches are equal to _____ half inch.

3. Travis says his yellow pencil measures $5\frac{1}{2}$ inches. Ralph says that is the same as 11 half inches. Explain how they are both correct.

EUREKA MATH

Lesson 5:    Create ruler with 1-inch, ½-inch, and ¼-inch intervals, and generate measurement data.

©2015 Great Minds. eureka-math.org
G3-M6-TE-B6-1.3.1-01.2016

75

lined paper

**Lesson 5:** Create ruler with 1-inch, ½-inch, and ¼-inch intervals, and generate measurement data.

©2015 Great Minds. eureka-math.org
G3-M6-TE-B6-1.3.1-01.2016

**EUREKA MATH**

# Lesson 6

Objective: Interpret measurement data from various line plots.

## Suggested Lesson Structure

- ■ Fluency Practice          (14 minutes)
- ▨ Application Problem     (5 minutes)
- ▢ Concept Development    (31 minutes)
- ■ Student Debrief           (10 minutes)

   **Total Time**               **(60 minutes)**

## Fluency Practice  (14 minutes)

- Group Counting  **3.OA.1**          (3 minutes)
- Multiply by 6  **3.OA.7**            (7 minutes)
- Read Bar Graphs  **3.MD.3**         (4 minutes)

## Group Counting  (3 minutes)

Note: Group counting reviews interpreting multiplication as repeated addition.

- T:   Count by sevens to 70.  (Write as students count.)
- S:   7, 14, 21, 28, 35, 42, 49, 56, 63, 70.
- T:   Let's count again.  Try not to look at the board.  When I raise my hand, stop.
- S:   7, 14, 21.
- T:   (Raise hand.)  21 is the same as how many sevens?
- S:   3 sevens.
- T:   Say 3 sevens as a multiplication sentence.
- S:   $3 \times 7 = 21$.
- T:   Continue.
- S:   28, 35, 42, 49, 56.
- T:   (Raise hand.)  56 is how many sevens?
- S:   8 sevens.
- T:   Say 8 sevens as a multiplication sentence.
- S:   $8 \times 7 = 56$.
- T:   (Write $14 \div 7 = $ ____.)  Let's find the answer counting by sevens.
- S:   7, 14.

T:   How many sevens are in 14?

S:   2 sevens.

T:   Say the division number sentence.

S:   $14 \div 7 = 2$.

Continue the process for the following possible sequence:  $28 \div 7$ and $63 \div 7$.

## Multiply by 6  (7 minutes)

Materials:   (S) Multiply by 6 (1–5) (Pattern Sheet)

Note:  This activity builds fluency with multiplication facts using units of 6.  It works toward students knowing from memory all products of two one-digit numbers.

T:   (Write $5 \times 6 =$ ____.)  Let's skip-count up by sixes to find the answer.  (Raise a finger for each number to track the count.  Record the skip-count answers on the board.)

S:   6, 12, 18, 24, 30.

T:   (Circle 30, and write $5 \times 6 = 30$ above it.  Write $3 \times 6 =$ ____.)  Let's skip-count up by sixes again.  (Track with fingers as students count.)

S:   6, 12, 18.

T:   Let's see how we can skip-count down to find the answer, too.  Start at 30 with 5 fingers, 1 for each six.  (Count down with your fingers as students say numbers.)

S:   30 (5 fingers), 24 (4 fingers), 18 (3 fingers).

Repeat the process for $4 \times 6$.

T:   (Distribute the Multiply by 6 Pattern Sheet.)  Let's practice multiplying by 6.  Be sure to work left to right across the page.

### Directions for Administration of Multiply-By Pattern Sheet

- Distribute the Multiply-By Pattern Sheet.
- Allow a maximum of two minutes for students to complete as many problems as possible.
- Direct students to work left to right across the page.
- Encourage skip-counting strategies to solve unknown facts.

## Read Bar Graphs  (4 minutes)

Materials:   (T) Number of Miles bar graph (Fluency Template) pictured next page  (S) Personal white board

Note:  This fluency activity reviews Lesson 4.  Students may initially need support beyond what is written below to find the exact number of miles driven, slightly extending the time this activity takes.

T:   (Project the bar graph.)  What does this bar graph show?

S:   The number of miles a truck driver drove Monday through Friday.

T:   On which day did the truck driver drive the most miles?

S:   Wednesday.

T:  On which day did the truck driver drive the least number of miles?

S:  Thursday.

T:  What is the scale for number of miles?

S:  50.

T:  How many intervals are there between each 50?

S:  5.

T:  On your boards, write a number sentence to show the value of the smaller intervals.

S:  (Write $50 \div 5 = 10$.)

T:  How many miles did the truck driver drive on Monday?

S:  340 miles.

T:  (Write 340 miles.)

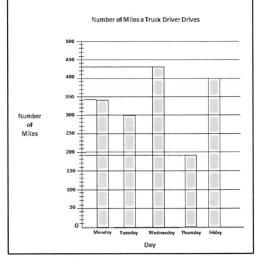

Continue the process for the following:  Tuesday, Wednesday, Thursday, and Friday.

T:  Write a number sentence to find how many miles the truck driver drove from Monday through Wednesday.

S:  (Write $340 + 300 + 430 = 1{,}070$.)

T:  Write a number sentence to find how many more miles the truck driver drove on Friday than on Thursday.

S:  (Write $400 - 190 = 210$.)

## Application Problem  (5 minutes)

Katelynn measures the height of her bean plant on Monday and again on Friday.  She says that her bean plant grew 10 quarter inches.  Her partner records $2\frac{1}{2}$ inches on his growth chart for the week.  Is her partner right?  Why or why not?

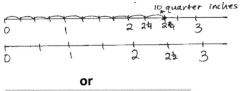

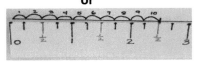

**or**

Yes, her partner is right. I drew a ruler divided into quarter inches and 10 quarter inches is $2\frac{2}{4}$ inches. Then I drew another ruler divided into half inches. I can see that $2\frac{2}{4}$ is the same as $2\frac{1}{2}$ on my rulers.

Note:  This problem reviews the relationship between quarter, half, and whole inches from Lesson 5. Students can choose to draw their own rulers or use the rulers they made from Lesson 5 to solve the problem.

## Concept Development (31 minutes)

**Materials:** (T) Time Spent Outside line plot (Template) pictured to the right  (S) Personal white board, blank paper, markers, Time Spent Outside line plot (Template)

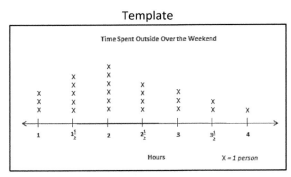

Template

**Problem 1: Use line plots with fractions to display measurement data.**

T: (Project the line plot, but only reveal the number line, as shown below. Point to the tick mark between 1 and 2.)  What should I label this tick mark on the number line?

S: $1\frac{1}{2}$ because it looks like it is halfway between 1 and 2.

T: (Label $1\frac{1}{2}$.)  When I point to each tick mark, tell me what to write.  (Point to the tick marks between 2 and 3 and then 3 and 4, labeling them $2\frac{1}{2}$ and $3\frac{1}{2}$, respectively.)

T: Talk to a partner.  How is this number line similar to the ruler we made yesterday?  How is it different?

S: They both show the numbers from 1 to 4.  The ruler actually goes to 6 inches. → They are both lines marked with whole units and fractional units. → The number line shows the halves between each whole number, but the ruler shows quarter inches too.

T: (Reveal the rest of the line plot.)  What does the number 1 on this line plot represent?

S: 1 hour.

T: What does the number $1\frac{1}{2}$ represent?

S: One and 1 half hours. → One full hour and half of another hour. → One hour and 30 minutes.

T: If the label on our line plot was people instead of hours, could we have fractions?

S: What is a fraction of a person? → My dad always says, "When I was half your size." → No, it would not make sense because you cannot have a fraction of a person.

T: So, when we use fractions on line plots, we need to make sure that it makes sense for the units to be given as fractions.  Talk to your partner.  What else besides time could you show on a line plot with fractions?

S: The lengths of our straws from yesterday. → The heights of our classmates. → Our shoe sizes. → The heights of our bean plants. → Anything we can measure!

T: That's right!  We can show measurements on a line plot with fractions.  How is a line plot like a bar graph or tape diagram?

S: The X's are like the units of 1 in a tape diagram. → The X's look like bars. → The tallest column of X's shows the most.

Lesson 6:          Interpret measurement data from various line plots.

T:   Which amount of time spent outside has the most X's?

S:   2 hours!

T:   When we made bar graphs and picture graphs, we used the word *favorite* to talk about the data that had the largest value.  Does it make sense to say 2 hours is the favorite amount of time spent outside?

S:   No.

T:   We can say that 2 hours is the most **frequent** or common amount of time spent outside because it has the most X's.  What is the second most frequent amount of time spent outside?

S:   $1\frac{1}{2}$ hours.

T:   What does each X on the line plot represent?

S:   A person!

T:   How many people spent $2\frac{1}{2}$ hours outside?

S:   4 people!

**NOTES ON
MULTIPLE MEANS
OF ENGAGEMENT:**

Give English language learners guided practice using *frequent, common, at least, more than,* and *less than* as they speak and write their observations about data.  Have students practice with partners using sentence frames like the ones below.

- The most frequently used word in our class is ____.

- ____ students read for at least 20 minutes last night.

- The most common excuse for not having homework is ____.

**Problem 2:  Read and interpret line plots with fractions.**

Students work in groups of four to write true statements about the Time Spent Outside line plot.  The goal is to write as many true statements as possible in the time given.  Each student in the group uses a different colored marker and can only write with his or her specified color.  This ensures engagement and equal participation in this activity.  Groups then prepare a poster with their statements to present to the class.

If time allows, the class can create a new line plot for this part of the lesson.  Students can measure their pencils to the nearest quarter inch.  Then, they can record their pencil's measurement on a class line plot, using stickers (e.g., stars or colored dots) or by making X's.

Prepare students:

1.   Write a list of words that the students must include in their statements.  This list should include the following words: *at least, frequent, less than,* and *more than.*  Be sure to check for understanding of these words.

2.   To achieve the highest score of 4, each of the following must be included and be correct:

   a.   A statement using the word *frequent* or *common.*
   b.   A statement using the words *at least.*
   c.   A comparison statement using *more than* requiring subtraction to solve.
   d.   A comparison statement using *less than* requiring subtraction to solve.

3.   Remind students that the amount of each color marker will be observed to check for equal participation.

For the presentation, students can do a modified gallery walk where one student from each group stays at the poster to be available to answer any questions about the statements.

## Problem Set  (10 minutes)

Students should do their personal best to complete the Problem Set within the allotted 10 minutes.  For some classes, it may be appropriate to modify the assignment by specifying which problems they work on first.  Some problems do not specify a method for solving.  Students should solve these problems using the RDW approach used for Application Problems.

## Student Debrief  (10 minutes)

**Lesson Objective:**  Interpret measurement data from various line plots.

The Student Debrief is intended to invite reflection and active processing of the total lesson experience.

Invite students to review their solutions for the Problem Set. They should check work by comparing answers with a partner before going over answers as a class.  Look for misconceptions or misunderstandings that can be addressed in the Debrief. Guide students in a conversation to debrief the Problem Set and process the lesson.

Any combination of the questions below may be used to lead the discussion.

- Using your answers from Problems 1(a) and (b), what subtraction sentence could you use to find the number of children who are at least 53 inches tall?  (15 − 6 = 9.)

- How many half inches does the child who is $52\frac{1}{2}$ inches tall need to grow to be tall enough to do the tip-off?

- What is the most **frequent** length of the worms in Problem 2?  How do you know?

- What kind of data can be shown on a line plot with fractions?  Are there any limitations?

- How did the Application Problem prepare you for today's lesson?

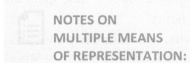

**NOTES ON MULTIPLE MEANS OF REPRESENTATION:**

Students working below grade level may benefit from modifications to Problem 2 of the Problem Set that make the data easier to discern. Consider the following:

- Enlarge the Lengths of Worms line plot.

- Have students label the column totals.

- Have students mark or highlight data they have counted.

- Draw rectangles around data in each column, or cover remaining data with a piece of paper to help students focus on one set of data at a time.

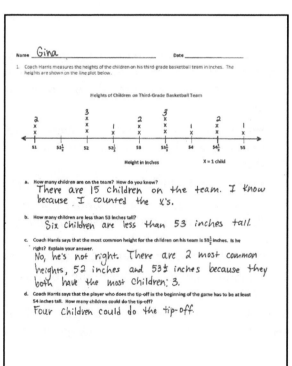

## Exit Ticket  (3 minutes)

After the Student Debrief, instruct students to complete the Exit Ticket.  A review of their work will help with assessing students' understanding of the concepts that were presented in today's lesson and planning more effectively for future lessons.  The questions may be read aloud to the students.

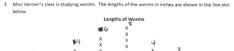
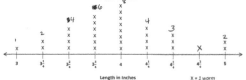

2.  Miss Vernier's class is studying worms.  The lengths of the worms in inches are shown in the line plot below.

**Lengths of Worms**

Length in inches        X = 1 worm

a.  How many worms did the class measure?  How do you know?

   The class measured 30 worms. I know because I counted the X's.

b.  Cara says that there are more worms $3\frac{3}{4}$ inches long than worms that are $3\frac{2}{4}$ and $4\frac{1}{4}$ inches long combined.  Is she right?  Explain your answer.

   6 worms are $3\frac{3}{4}$ inches.
   $4+4=8$ worms that are $3\frac{2}{4}$ and $4\frac{1}{4}$ inches.
   No, she's wrong because there are more worms that are $3\frac{2}{4}$ long and $4\frac{1}{4}$ long than $3\frac{3}{4}$ inches long.

c.  Madeline finds a worm hiding under a leaf.  She measures it, and it is $4\frac{3}{4}$ inches long.  Plot the length of the worm on the line plot.

Multiply.

6 x 1 = _____    6 x 2 = _____    6 x 3 = _____    6 x 4 = _____

6 x 5 = _____    6 x 1 = _____    6 x 2 = _____    6 x 1 = _____

6 x 3 = _____    6 x 1 = _____    6 x 4 = _____    6 x 1 = _____

6 x 5 = _____    6 x 1 = _____    6 x 2 = _____    6 x 3 = _____

6 x 2 = _____    6 x 4 = _____    6 x 2 = _____    6 x 5 = _____

6 x 2 = _____    6 x 1 = _____    6 x 2 = _____    6 x 3 = _____

6 x 1 = _____    6 x 3 = _____    6 x 2 = _____    6 x 3 = _____

6 x 4 = _____    6 x 3 = _____    6 x 5 = _____    6 x 3 = _____

6 x 4 = _____    6 x 1 = _____    6 x 4 = _____    6 x 2 = _____

6 x 4 = _____    6 x 3 = _____    6 x 4 = _____    6 x 5 = _____

6 x 4 = _____    6 x 5 = _____    6 x 1 = _____    6 x 5 = _____

6 x 2 = _____    6 x 5 = _____    6 x 3 = _____    6 x 5 = _____

6 x 4 = _____    6 x 2 = _____    6 x 4 = _____    6 x 3 = _____

6 x 5 = _____    6 x 3 = _____    6 x 2 = _____    6 x 4 = _____

6 x 3 = _____    6 x 5 = _____    6 x 2 = _____    6 x 4 = _____

multiply by 6 (1–5)

**Lesson 6:**    Interpret measurement data from various line plots.

EUREKA MATH™

Name _____　　Date _____

1.　Coach Harris measures the heights of the children on his third-grade basketball team in inches. The heights are shown on the line plot below.

**Heights of Children on Third-Grade Basketball Team**

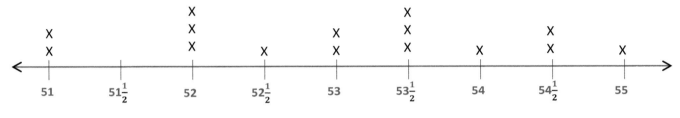

**Height in Inches**　　　　　　　　　　　X = 1 child

a.　How many children are on the team? How do you know?

b.　How many children are less than 53 inches tall?

c.　Coach Harris says that the most common height for the children on his team is $53\frac{1}{2}$ inches. Is he right? Explain your answer.

d.　Coach Harris says that the player who does the tip-off in the beginning of the game has to be at least 54 inches tall. How many children could do the tip-off?

2. Miss Vernier's class is studying worms. The lengths of the worms in inches are shown in the line plot below.

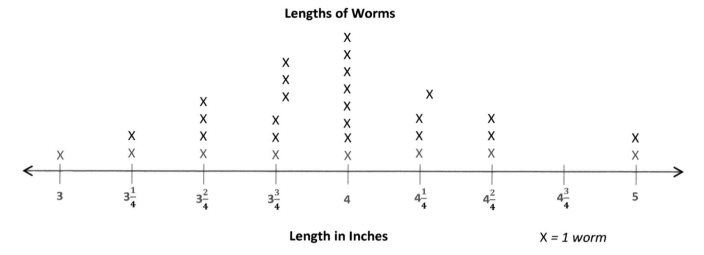

**Lengths of Worms**

**Length in Inches**

X = 1 worm

a. How many worms did the class measure? How do you know?

b. Cara says that there are more worms $3\frac{3}{4}$ inches long than worms that are $3\frac{2}{4}$ and $4\frac{1}{4}$ inches long combined. Is she right? Explain your answer.

c. Madeline finds a worm hiding under a leaf. She measures it, and it is $4\frac{3}{4}$ inches long. Plot the length of the worm on the line plot.

EUREKA MATH

Name _____   Date _____

Ms. Bravo measures the lengths of her third-grade students' hands in inches.  The lengths are shown on the line plot below.

**Lengths of Hands of Third-Grade Students**

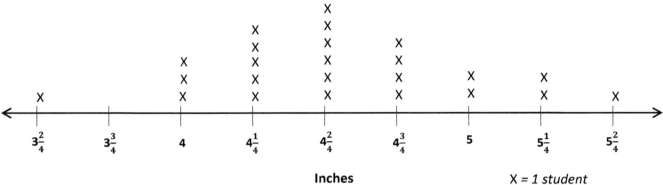

**Inches**                              X = 1 student

a.  How many students are in Ms. Bravo's class?  How do you know?

b.  How many students' hands are longer than $4\frac{2}{4}$ inches?

c.  Darren says that more students' hands are $4\frac{2}{4}$ inches long than 4 and $5\frac{1}{4}$ inches combined.  Is he right?  Explain your answer.

Name _____     Date _____

1.  Ms. Leal measures the heights of the students in her kindergarten class.  The heights are shown on the line plot below.

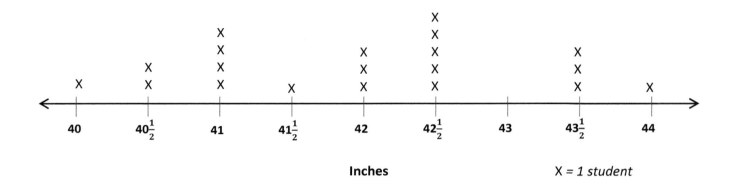

**Heights of Students in Ms. Leal's Kindergarten Class**

Inches                                      X = 1 student

a.  How many students in Ms. Leal's class are exactly 41 inches tall?

b.  How many students are in Ms. Leal's class?  How do you know?

c.  How many students in Ms. Leal's class are more than 42 inches tall?

d.  Ms. Leal says that for the class picture students in the back row must be at least $42\frac{1}{2}$ inches tall.  How many students should be in the back row?

**Lesson 6:**    Interpret measurement data from various line plots.

EUREKA
MATH™

2. Mr. Stein's class is studying plants. They plant seeds in clear plastic bags and measure the lengths of the roots. The lengths of the roots in inches are shown in the line plot below.

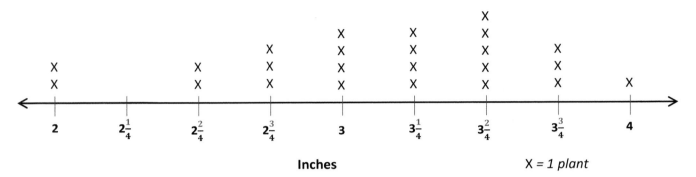

**Lengths of Plants' Roots**

**Inches**  X = 1 plant

a. How many roots did Mr. Stein's class measure? How do you know?

b. Teresa says that the 3 most frequent measurements in order from shortest to longest are $3\frac{1}{4}$ inches, $3\frac{2}{4}$ inches, and $3\frac{3}{4}$ inches. Do you agree? Explain your answer.

c. Gerald says that the most common measurement is 14 quarter inches. Is he right? Why or why not?

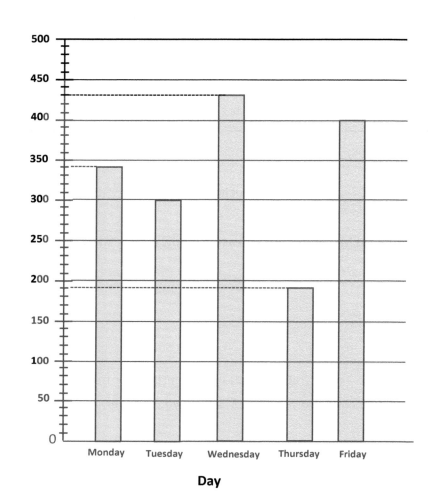

**Number of Miles a Truck Driver Drives**

number of miles bar graph

**Lesson 6:**        Interpret measurement data from various line plots.

**EUREKA MATH™**

**Time Spent Outside Over the Weekend**

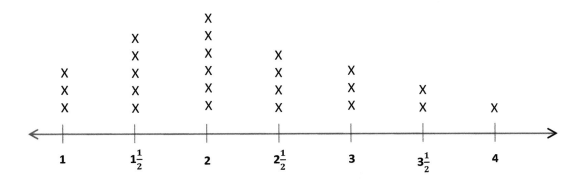

**Hours**                    X = 1 person

_____

time spent outside line plot

# Lesson 7

Objective:  Represent measurement data with line plots.

## Suggested Lesson Structure

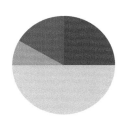

■ Fluency Practice          (15 minutes)
■ Application Problem        (5 minutes)
■ Concept Development        (30 minutes)
■ Student Debrief           (10 minutes)

   **Total Time**            **(60 minutes)**

## Fluency Practice  (15 minutes)

- Group Counting  **3.OA.1**                    (3 minutes)
- Multiply by 6  **3.OA.7**                     (8 minutes)
- Count by Halves and Fourths  **3.MD.4**       (4 minutes)

### Group Counting  (3 minutes)

Note:  This group counting activity reviews the relationship between counting by a unit and multiplying and dividing with that unit.

   T:   Count by sevens to 70.
   S:   7, 14, 21, 28, 35, 42, 49, 56, 63, 70.
   T:   (Write $4 \times 7 =$ ___.)  What is the value of 4 sevens?  Count by sevens if you are unsure.
   S:   28.
   T:   Say the multiplication sentence.
   S:   $4 \times 7 = 28$.

Continue this process for $6 \times 7$ and $8 \times 7$.

   T:   (Write $21 \div 7 =$ ___.)  What is $21 \div 7$?  Count by sevens if you are unsure.
   S:   3.

Continue this process for $35 \div 7$, $49 \div 7$, and $63 \div 7$.

   T:   Count by eights to 80.
   S:   8, 16, 24, 32, 40, 48, 56, 64, 72, 80.
   T:   (Write $3 \times 8 =$ ___.)  What is the value of 3 eights?
   S:   24.

EUREKA
MATH™

T:   Say the multiplication sentence.

S:   3 × 8 = 24.

Continue this process for 6 × 8 and 8 × 8.

T:   (Write 24 ÷ 8 = ___.) What is 24 ÷ 8?  Count by eights if you are unsure.

S:   3.

Continue this process for 32 ÷ 8, 56 ÷ 8, and 72 ÷ 8.

## Multiply by 6  (8 minutes)

Materials:   (S) Multiply by 6 (6–10) (Pattern Sheet)

Note:  This activity builds fluency with multiplication facts using units of 6.  It works toward students knowing from memory all products of two one-digit numbers.  See Lesson 6 for the directions for administration of a Multiply-By Pattern Sheet.

T:   (Write 7 × 6 = ___.) Let's skip-count up by sixes.  I'll raise a finger for each six.  (Raise a finger for each number to track the count.  Record the skip-count answers on the board.)

S:   6, 12, 18, 24, 30, 36, 42.

T:   Let's see how we can skip-count down to find the answer, too.  Start at 60 with 10 fingers, 1 for each six.  (Count down with fingers as students say numbers.)

S:   60 (10 fingers), 54 (9 fingers), 48 (8 fingers), 42 (7 fingers).

Continue with the following suggested sequence:  9 × 6, 6 × 6, and 8 × 6.

T:   (Distribute the Multiply by 6 Pattern Sheet.)  Let's practice multiplying by 6.  Be sure to work left to right across the page.

## Count by Halves and Fourths  (4 minutes)

Note:  This activity reviews Lesson 6.

T:   Count by halves to 12 halves as I write.  Please do not count faster than I can write.  (Write in fractional form as students count.)

S:   1 half, 2 halves, 3 halves, ..., 11 halves, 12 halves.

T:   (Point to $\frac{2}{2}$.)  Say 2 halves as a whole number.

S:   1.

T:   (Lightly cross out $\frac{2}{2}$, and write 1 beneath it.)

Halves:

$$\frac{1}{2} \quad \frac{2}{2} \quad \frac{3}{2} \quad \frac{4}{2} \quad \frac{5}{2} \quad \frac{6}{2} \quad \frac{7}{2} \quad \frac{8}{2} \quad \frac{9}{2} \quad \frac{10}{2} \quad \frac{11}{2} \quad \frac{12}{2}$$
$$\quad 1 \qquad 2 \qquad 3 \qquad 4 \qquad 5 \qquad 6$$

Fourths:

$$\frac{1}{4} \quad \frac{2}{4} \quad \frac{3}{4} \quad \frac{4}{4} \quad \frac{5}{4} \quad \frac{6}{4} \quad \frac{7}{4} \quad \frac{8}{4} \quad \frac{9}{4} \quad \frac{10}{4} \quad \frac{11}{4} \quad \frac{12}{4}$$
$$\qquad\qquad 1 \qquad\qquad 2 \qquad\qquad 3$$

Continue the process for the following sequence:  $\frac{4}{2}, \frac{6}{2}, \frac{8}{2}, \frac{10}{2}$, and $\frac{12}{2}$.

T:   Count by halves.  Say whole numbers when you arrive at whole numbers.  Try not to look at the board.  (Students count forward and backward on the number line.  Occasionally change directions.)

Repeat the process for fourths.

## Application Problem  (5 minutes)

The chart shows the lengths of straws measured in Mr. Han's class.

a.  How many straws were measured?  Explain how you know.

b.  What is the smallest measurement on the chart?  The greatest?

c.  Were the straws measured to the nearest inch?  How do you know?

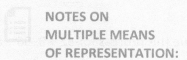

**NOTES ON MULTIPLE MEANS OF REPRESENTATION:**

Use color to customize the presentation of the data in the chart. Enhance learners' perception of the information by lightly shading every other row, highlighting numerators and/or denominators, or consistently writing whole numbers in a specific color (e.g., using red for 4).

| Straw Lengths (in Inches) | | | | |
|---|---|---|---|---|
| 3 | 4 | $4\frac{1}{2}$ | $2\frac{3}{4}$ | $3\frac{3}{4}$ |
| $3\frac{3}{4}$ | $4\frac{1}{2}$ | $3\frac{1}{4}$ | 4 | $4\frac{3}{4}$ |
| $4\frac{1}{4}$ | 5 | 3 | $3\frac{1}{2}$ | $4\frac{1}{2}$ |
| $4\frac{1}{2}$ | 4 | $3\frac{1}{4}$ | 5 | $4\frac{1}{4}$ |

a) 20 straws were measured. I know this because there are 20 measurements on the chart, and each represents 1 straw.

b) The smallest measurement is $2\frac{3}{4}$ inches. The greatest measurement is 5 inches.

c) No, they weren't measured to the nearest inch because there are also quarter-inches and half-inches.

Note:  The Straw Lengths chart is included on the template used in the Concept Development.  Rather than recreate it for this problem, the template can be projected instead.  Students use the measurements from the chart to create a line plot in the Concept Development.  The questions from the Application Problem help facilitate the discussion in the Concept Development about how to create a scale for the line plot.

## Concept Development  (30 minutes)

Materials:  (S) Student-made ruler from Lesson 5, Straw Lengths (Template) pictured with Problem 2 below

**Problem 1: Draw a line plot representing measurement data.**

T:  Let's represent the straw data from Mr. Han's class using a line plot.  First, we need to determine the scale for our line plot.  The first measurement on the line plot should be the smallest measurement in the chart.  What is the smallest measurement?

S:  $2\frac{3}{4}$ inches.

T: What do you think will be the last measurement on the line plot?

S: 5 inches because it is the largest measurement.

T: Turn and talk to your partner. Look over the data in the chart. How do you know what interval we should count by to create our scale?

S: Counting by whole inches is the easiest, but it does not allow us to plot all of our numbers. → The data has numbers with whole inches, half inches, and quarter inches. It makes the most sense to count by quarter inches because they are the smallest.

T: To find out how many tick marks we need, we can count by fourths from $2\frac{3}{4}$ to 5. Each time we count, keep track with your fingers.

T: Let's count.

S: (Track the count by fourths from $2\frac{3}{4}$ to 5.)

T: How many tick marks do we need to draw altogether?

S: 10 tick marks.

**MP.6**

T: I heard some count $3\frac{2}{4}$ and others count $3\frac{1}{2}$. Who is correct? Talk to your partner.

S: 2 fourths equals a half, so they are the same. → $\frac{1}{4}$ and $\frac{1}{4}$ is the same as $\frac{1}{2}$.

T: Both fractions name the same length. In the data chart, it is written as $3\frac{1}{2}$, so it is best to label it the same way.

T: (Pass out the template). On the template, you see the chart from the Application Problem and an empty number line. We need to partition our number line into equal intervals and label our scale. How can we use our ruler to create equal intervals?

S: We can make a mark at every inch until we have 10 marks.

T: Draw to show 10 marks. Then, label each mark from $2\frac{3}{4}$ to 5 inches. (Model as students work.)

S: (Draw and label.)

**Problem 2: Plot data set on the line plot.**

T: Now, it is time to record the data on our line plot. Look at the first measurement in the chart. Look for that measurement on your line plot. (Allow time for students to locate.)

T: Plot that data on the line plot with an X. (Model.) How can we make sure that we plot the data only once?

S: We can cross or check off each one as we go.

T: Plot the rest of the data with care, either crossing or checking off each measure you plot. (Allow students time to work.)

T: Let's give this line plot a title that tells what it shows. What data is represented on the line plot?

S: Lengths of different straws.

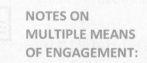

**NOTES ON MULTIPLE MEANS OF ENGAGEMENT:**

Encourage students working below grade level and others to whisper-read the data as they plot if this helps them track the information. Students may work in pairs.

Alternatively, challenge students working above grade level to offer two other representations of the data (e.g., picture graph, bar graph, tape diagram, tally chart). Have students compare and list the advantages of using a line plot.

EUREKA MATH™

T:   Let's title our line plot *Straw Lengths*. (Model.) Add the title to your graph. (Allow students time to work.) Let's add a key to show what each X represents. What does each X represent?

S:   A straw!

T:   (Model adding a key to the line plot.) Add a key to your line plot. (Allow students time to work.) Let's also put a label beneath the number line to tell the unit our line plot shows. What unit did we use to measure?

S:   Inches!

T:   Let's add the word *Inches* underneath the numbers on the number line. (Model.) Now that our line plot has a title, a key, and a unit label, anybody who looks at the line plot will know what it is showing.

Continue with the following suggested questions:

- How many straws were at least ___ inches tall?
- How many straws were taller/shorter than ___ inches?
- Which measurements happened most/least frequently?

Sample Work

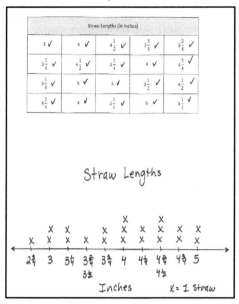

## Problem Set  (10 minutes)

Students should do their personal best to complete the Problem Set within the allotted 10 minutes. For some classes, it may be appropriate to modify the assignment by specifying which problems they work on first. Some problems do not specify a method for solving. Students should solve these problems using the RDW approach used for Application Problems.

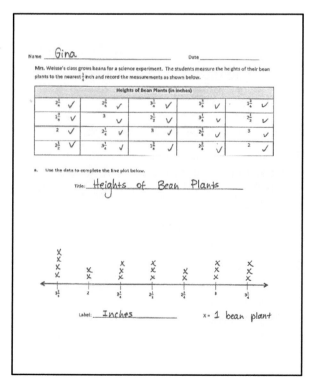

## Student Debrief  (10 minutes)

**Lesson Objective:** Represent measurement data with line plots.

The Student Debrief is intended to invite reflection and active processing of the total lesson experience.

Invite students to review their solutions for the Problem Set. They should check work by comparing answers with a partner before going over answers as a class. Look for misconceptions or misunderstandings that can be addressed in the Debrief. Guide students in a conversation to debrief the Problem Set and process the lesson.

EUREKA
MATH™

Any combination of the questions below may be used to lead the discussion.

- What process did you use to complete the line plot in Problem (a)?
- What other questions could be answered based on the Heights of Bean Plants data?
- Why do you think four of the bean plants were so short?  What questions would you ask Mrs. Weisse's class about this?  (Possible student answers:  Did they have different soil?  Were they short but very healthy?  Was it a different kind of bean plant?)
- In what ways is a line plot similar to a picture graph in how it displays data?  Bar graph?  In what ways is it different?
- Why is it important to create a scale before partitioning a number line?
- In what ways did your knowledge of fractions help you create your line plots?
- How did the Fluency Practice activities connect to today's new learning?
- How did the Application Problem help you get ready for today's lesson?

b.  How many bean plants are at least 2¼ inches tall?

14 bean plants are at least 2¼ inches tall.

c.  How many bean plants are taller than 2¾ inches?

6 bean plants are taller than 2¾ inches.

d.  What is the most frequent measurement?  How many bean plants were plotted for this measurement?

The most frequent measurement is 1¾ inches. 4 bean plants were plotted for that measurement.

e.  George says that most of the bean plants are at least 3 inches tall.  Is he right?  Explain your answer.

George is not right. Only 6 bean plants are at least 3 inches tall. 14 bean plants are shorter than 3 inches. 14 is more than 6.

f.  Savannah was absent the day the class measured the height of their bean plants.  When she returns, her plant measures 2¾ inches tall.  Can Savannah plot the height of her bean plant on the class line plot?  Why or why not?

Yes, Savannah can plot the height of her bean plant. 2 4/8 is the same as 2½, so she can draw an 'X' at 2½ inches.

## Exit Ticket  (3 minutes)

After the Student Debrief, instruct students to complete the Exit Ticket.  A review of their work will help with assessing students' understanding of the concepts that were presented in today's lesson and planning more effectively for future lessons.  The questions may be read aloud to the students.

Multiply.

6 x 1 = _____     6 x 2 = _____     6 x 3 = _____     6 x 4 = _____

6 x 5 = _____     6 x 6 = _____     6 x 7 = _____     6 x 8 = _____

6 x 9 = _____     6 x 10 = _____     6 x 5 = _____     6 x 6 = _____

6 x 5 = _____     6 x 7 = _____     6 x 5 = _____     6 x 8 = _____

6 x 5 = _____     6 x 9 = _____     6 x 5 = _____     6 x 10 = _____

6 x 6 = _____     6 x 5 = _____     6 x 6 = _____     6 x 7 = _____

6 x 6 = _____     6 x 8 = _____     6 x 6 = _____     6 x 9 = _____

6 x 6 = _____     6 x 7 = _____     6 x 6 = _____     6 x 7 = _____

6 x 8 = _____     6 x 7 = _____     6 x 9 = _____     6 x 7 = _____

6 x 8 = _____     6 x 6 = _____     6 x 8 = _____     6 x 7 = _____

6 x 8 = _____     6 x 9 = _____     6 x 9 = _____     6 x 6 = _____

6 x 9 = _____     6 x 7 = _____     6 x 9 = _____     6 x 8 = _____

6 x 9 = _____     6 x 8 = _____     6 x 6 = _____     6 x 9 = _____

6 x 7 = _____     6 x 9 = _____     6 x 6 = _____     6 x 8 = _____

6 x 9 = _____     6 x 7 = _____     6 x 6 = _____     6 x 8 = _____

multiply by 6 (6–10)

    **Lesson 7:**     Represent measurement data with line plots.

**EUREKA MATH**

Name _____    Date _____

Mrs. Weisse's class grows beans for a science experiment.  The students measure the heights of their bean plants to the nearest $\frac{1}{4}$ inch and record the measurements as shown below.

| Heights of Bean Plants (in Inches) | | | | |
|---|---|---|---|---|
| $2\frac{1}{4}$ | $2\frac{3}{4}$ | $3\frac{1}{4}$ | $1\frac{3}{4}$ | $1\frac{3}{4}$ |
| $1\frac{3}{4}$ | $3$ | $2\frac{1}{2}$ | $3\frac{1}{4}$ | $2\frac{1}{2}$ |
| $2$ | $2\frac{1}{4}$ | $3$ | $2\frac{1}{4}$ | $3$ |
| $2\frac{1}{2}$ | $3\frac{1}{4}$ | $1\frac{3}{4}$ | $2\frac{3}{4}$ | $2$ |

a.  Use the data to complete the line plot below.

    Title: _____

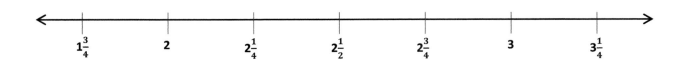

    Label: _____    X =

b.  How many bean plants are at least $2\frac{1}{4}$ inches tall?

c.  How many bean plants are taller than $2\frac{3}{4}$ inches?

d.  What is the most frequent measurement?  How many bean plants were plotted for this measurement?

e.  George says that most of the bean plants are at least 3 inches tall.  Is he right?  Explain your answer.

f.  Savannah was absent the day the class measured the heights of their bean plants.  When she returns, her plant measures $2\frac{2}{4}$ inches tall.  Can Savannah plot the height of her bean plant on the class line plot?  Why or why not?

Lesson 7:     Represent measurement data with line plots.

©2015 Great Minds. eureka-math.org
G3-M6-TE-B6-1.3.1-01.2016

Name _____     Date _____

Scientists measure the growth of mice in inches.  The scientists measure the length of the mice to the nearest $\frac{1}{4}$ inch and record the measurements as shown below.

| Lengths of Mice (in Inches) | | | | |
|---|---|---|---|---|
| $3\frac{1}{4}$ | $3$ | $3\frac{1}{4}$ | $3\frac{3}{4}$ | $4$ |
| $3\frac{3}{4}$ | $3$ | $4\frac{1}{2}$ | $4\frac{1}{2}$ | $3\frac{3}{4}$ |
| $4$ | $4\frac{1}{4}$ | $4$ | $4\frac{1}{4}$ | $4$ |

Label each tick mark.  Then, record the data on the line plot below.

Title: _____

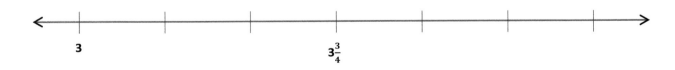

3                                                            $3\frac{3}{4}$

Label: _____ X = 1 mouse

EUREKA
MATH

Lesson 7:     Represent measurement data with line plots.

101

©2015 Great Minds. eureka-math.org
G3-M6-TE-B6-1.3.1-01.2016

Name _____     Date _____

Mrs. Felter's students build a model of their school's neighborhood out of blocks.  The students measure the heights of the buildings to the nearest $\frac{1}{4}$ inch and record the measurements as shown below.

| Heights of Buildings (in Inches) | | | | |
|---|---|---|---|---|
| $3\frac{1}{4}$ | $3\frac{3}{4}$ | $4\frac{1}{4}$ | $4\frac{1}{2}$ | $3\frac{1}{2}$ |
| 4 | 3 | $3\frac{3}{4}$ | 3 | $4\frac{1}{2}$ |
| 3 | $3\frac{1}{2}$ | $3\frac{3}{4}$ | $3\frac{1}{2}$ | 4 |
| $3\frac{1}{2}$ | $3\frac{1}{4}$ | $3\frac{1}{2}$ | 4 | $3\frac{3}{4}$ |
| 3 | $4\frac{1}{4}$ | 4 | $3\frac{1}{4}$ | 4 |

a.   Use the data to complete the line plot below.

Title: _____

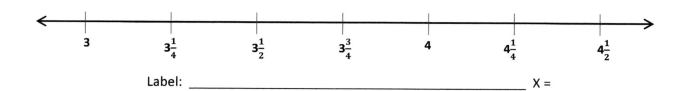

Label: _____  X =

**Lesson 7:**        Represent measurement data with line plots.

©2015 Great Minds. eureka-math.org
G3-M6-TE-B6-1.3.1-01.2016

b.  How many buildings are $4\frac{1}{4}$ inches tall?

c.  How many buildings are less than $3\frac{1}{2}$ inches?

d.  How many buildings are in the class model?  How do you know?

e.  Brook says most buildings in the model are at least 4 inches tall.  Is she correct?  Explain your thinking.

| Straw Lengths (in Inches) | | | | |
|:---:|:---:|:---:|:---:|:---:|
| 3 | 4 | $4\frac{1}{2}$ | $2\frac{3}{4}$ | $3\frac{3}{4}$ |
| $3\frac{3}{4}$ | $4\frac{1}{2}$ | $3\frac{1}{4}$ | 4 | $4\frac{3}{4}$ |
| $4\frac{1}{4}$ | 5 | 3 | $3\frac{1}{2}$ | $4\frac{1}{2}$ |
| $4\frac{3}{4}$ | 4 | $3\frac{1}{4}$ | 5 | $4\frac{1}{4}$ |

⟵⎯⎯⎯⎯⎯⎯⎯⎯⎯⎯⎯⎯⎯⎯⎯⎯⎯⎯⎯⎯⎯⟶

straw lengths

    **Lesson 7:**     Represent measurement data with line plots.

                                           **EUREKA MATH**

# Lesson 8

Objective:  Represent measurement data with line plots.

## Suggested Lesson Structure

■ Fluency Practice          (14 minutes)
▨ Application Problem        (3 minutes)
▧ Concept Development        (33 minutes)
■ Student Debrief           (10 minutes)

**Total Time**              **(60 minutes)**

## Fluency Practice  (14 minutes)

- Group Counting  **3.OA.1**                    (3 minutes)
- Multiply by 7  **3.OA.7**                     (7 minutes)
- Count by Halves and Fourths  **3.MD.4**       (4 minutes)

### Group Counting  (3 minutes)

Note:  This group counting activity reviews the relationship between counting by a unit and multiplying and dividing with that unit.

    T:  Count by eights to 80.
    S:  8, 16, 24, 32, 40, 48, 56, 64, 72, 80.
    T:  (Write $4 \times 8 =$ ___.)  What is the value of 4 eights?  Count by eights if you are unsure.
    S:  32.
    T:  Say the multiplication sentence.
    S:  $4 \times 8 = 32$.

Continue the process for $7 \times 8$ and $9 \times 8$.

    T:  (Write $24 \div 8 =$ ___.)  What is $24 \div 8$?  Count by eights if you are unsure.
    S:  3.

Continue the process for $40 \div 8$, $48 \div 8$, and $64 \div 8$.

    T:  Count by nines to 90.
    S:  9, 18, 27, 36, 45, 54, 63, 72, 81, 90.
    T:  (Write $2 \times 9 =$ ___.)  What is the value of 2 nines?  Count by nines if you are unsure.
    S:  18.

T:   Say the multiplication sentence.

S:   2 × 9 = 18.

Continue the process for 4 × 9, 6 × 9, and 8 × 9.

T:   (Write 27 ÷ 9 = ___.)  What is 27 ÷ 9?  Count by nines if you are unsure.

S:   3.

Continue the process for 45 ÷ 9, 63 ÷ 9, and 81 ÷ 9.

## Multiply by 7  (7 minutes)

Materials:   (S) Multiply by 7 (1–5) (Pattern Sheet)

Note:  This activity builds fluency with multiplication facts using units of 7.  It works toward students knowing from memory all products of two one-digit numbers.  See Lesson 6 for the directions for administration of a Multiply-By Pattern Sheet.

T:   (Write 5 × 7 = ___.)  Let's skip-count up by sevens to find the answer.  I'll raise a finger for each seven.  (Raise a finger for each number to track the count.  Record the skip-count answers on the board.)

S:   7, 14, 21, 28, 35.

T:   (Circle 35 and write 5 × 7 = 35 above it.  Write 3 × 7 = ___.)  Let's skip-count up by sevens again.  (Track with fingers as students count.)

S:   7, 14, 21.

T:   Let's see how we can skip-count down to find the answer, too.  Start at 35 with 5 fingers, 1 for each seven.  (Count down with your fingers as students say numbers.)

S:   35 (5 fingers), 28 (4 fingers), 21 (3 fingers).

Repeat the process for 4 × 7.

T:   (Distribute the Multiply by 7 Pattern Sheet.)  Let's practice multiplying by 7.  Be sure to work left to right across the page.

## Count by Halves and Fourths  (4 minutes)

Note:  This fluency activity reviews Lesson 6.

T:   Count by halves to 12 halves as I write.  Please do not count faster than I can write.  (Write as students count.)

S:   1 half, 2 halves, 3 halves, 4 halves, 5 halves, 6 halves, 7 halves, 8 halves, 9 halves, 10 halves, 11 halves, 12 halves.

T:   (Point to $\frac{2}{2}$.)  Say 2 halves as a whole number.

S:   1.

T:   (Lightly cross out $\frac{2}{2}$, and write 1 beneath it.)

Halves:

$$\frac{1}{2} \quad \frac{2}{2} \quad \frac{3}{2} \quad \frac{4}{2} \quad \frac{5}{2} \quad \frac{6}{2} \quad \frac{7}{2} \quad \frac{8}{2} \quad \frac{9}{2} \quad \frac{10}{2} \quad \frac{11}{2} \quad \frac{12}{2}$$

$$1 \qquad\quad 2 \qquad\quad 3 \qquad\quad 4 \qquad\quad 5 \qquad\quad 6$$

Fourths:

$$\frac{1}{4} \quad \frac{2}{4} \quad \frac{3}{4} \quad \frac{4}{4} \quad \frac{5}{4} \quad \frac{6}{4} \quad \frac{7}{4} \quad \frac{8}{4} \quad \frac{9}{4} \quad \frac{10}{4} \quad \frac{11}{4} \quad \frac{12}{4}$$

$$1 \qquad\qquad 2 \qquad\qquad 3$$

EUREKA
MATH™

Continue the process for the following sequence: $\frac{4}{2}, \frac{6}{2}, \frac{8}{2}, \frac{10}{2},$ and $\frac{12}{2}$.

T: Count by halves, saying whole numbers when you arrive at whole numbers. Try not to look at the board. (Direct students to count forward and backward on the number line, occasionally changing directions.)

Repeat the process for fourths.

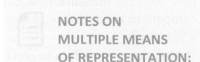

**NOTES ON MULTIPLE MEANS OF REPRESENTATION:**

Although the term *interval* was introduced in Module 2 and has been used earlier in this module, it may be appropriate to revisit its meaning for English language learners and others. Use drawings, gestures, and examples to explain the meaning of *interval*. Offer explanations in students' first language, if possible. Link vocabulary to synonyms they may be more familiar with, such as *space, period, distance,* and *gap* (on the number line).

## Application Problem (3 minutes)

Mrs. Byrne's class is studying worms. They measure the lengths of the worms to the nearest quarter inch. The length of the shortest worm is $3\frac{3}{4}$ inches. The length of the longest worm is $5\frac{2}{4}$ inches. Kathleen says they need 8 quarter-inch intervals to plot the lengths of the worms on a line plot. Is she right? Why or why not?

No, Kathleen is not right because they will need 7 quarter-inch intervals, not 8.

Note: This problem reviews Lesson 7, specifically using a quarter-inch scale to create a line plot. Invite students to discuss what Kathleen did wrong in her calculations. (She counted the numbers, not the intervals.) This problem provides an opportunity to discuss the number of tick marks versus the number of intervals.

## Concept Development (33 minutes)

Materials: (S) Heights of Sunflower Plants chart (Template) pictured to the right, personal white board, straightedge

**Problem 1: Plot a large data set to the nearest half inch.**

Students start with the Heights of Sunflower Plants Template in their personal white boards.

T: What data is shown in the chart?

S: The heights of sunflower plants.

Template

Mrs. Schaut measures the heights of the sunflower plants in her garden. The measurements are shown in the chart below.

| Heights of Sunflower Plants (in inches) | | | | |
|---|---|---|---|---|
| 61 | 63 | 62 | 61 | $62\frac{1}{2}$ |
| $61\frac{1}{2}$ | $61\frac{1}{2}$ | $61\frac{1}{2}$ | 62 | 60 |
| 64 | 62 | $60\frac{1}{2}$ | $63\frac{1}{2}$ | 61 |
| 63 | $62\frac{1}{2}$ | $62\frac{1}{2}$ | 64 | $62\frac{1}{2}$ |
| $62\frac{1}{2}$ | $63\frac{1}{2}$ | 63 | $62\frac{1}{2}$ | $63\frac{1}{2}$ |
| 62 | $62\frac{1}{2}$ | 62 | 63 | $60\frac{1}{2}$ |

Lesson 8: Represent measurement data with line plots.

107

T:    How does the **measurement data** in this chart compare to the measurement data we plotted yesterday?

S:    There is a lot more data to plot!  → The numbers are bigger too!

T:    Let's make a line plot to display it.  With a partner, discuss the steps you should take to create a line plot of the data.

S:    (Discuss.)

T:    What number does the first tick mark on your line plot represent?  How do you know?

S:    60 inches because it is the smallest measurement.

T:    And the last tick mark?  How do you know?

S:    64 inches because it is the biggest measurement.

T:    What interval should you use to draw the tick marks between 60 and 64?  How do you know?

S:    Half inches because that is what a lot of the measurements are.  → I should use half inches because it is a common unit in the chart.  → Half inches because it is the smallest unit in the chart.

T:    Go ahead and create your line plot.  (Circulate to check student work.)

NOTES ON
MULTIPLE MEANS
OF REPRESENTATION:

Give explicit prompts to students working below grade level for each step in the process of making a line plot for the Heights of Sunflower Plants data.  Make a poster, or speak the following:

- Find and record the smallest and largest measurements as endpoints.

- Choose the scale.  Ask, "What interval should I use—whole numbers, halves, or quarters?"

- Count by half inches from the smallest measurement to the largest measurement to find the number of tick marks to draw. Draw.

- Plot the data on the line plot. Check off each point along the way.

- Title the line plot (e.g., Heights of Sunflower Plants), and specify the units (e.g., inches).

**Problem 2:  Observe and interpret data on a line plot.**

T:    Tell me a true statement about the heights of the sunflower plants in Mrs. Schaut's garden.

S:    The most common height is $62\frac{1}{2}$ inches.  → There is only 1 plant that is 60 inches tall.  → 61, $61\frac{1}{2}$, and $63\frac{1}{2}$ inches all have the same number of plants.  → There are more plants that are $62\frac{1}{2}$ inches tall than 60, $60\frac{1}{2}$, and 61 inches combined.

T:    Are these statements true of the data in the chart?

S:    Yes because it is the same data.  We just displayed it differently.

T:    How does having the data displayed as a line plot help you to think and talk about the data?

S:    I can easily see the number of plants for each measurement.  → I can quickly see the most common and least common measurements.

T:    What are the three most frequent measurements in order from shortest to tallest?

S:    62, $62\frac{1}{2}$, and 63 inches.

T:    What is the total number of plants that measure 62, $62\frac{1}{2}$, and 63 inches?

S:    16 plants!

T: How many plants were measured in all?

S: 30 plants.

T: Write a number sentence to show how many plants do not measure 62, $62\frac{1}{2}$, or 63 inches.

S: (Write 30 − 16 = 14.)

T: (Write or say, "Most of the sunflower plants measure between 62 and 63 inches.")  True?

S: Yes! → Yes because 16 plants measure between 62 and 63 inches, and 14 plants do not.  Sixteen is more than 14.

T: What do you notice about the location of the three most frequent measurements on the line plot?

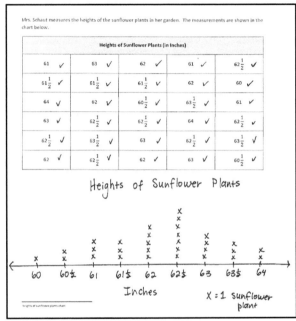

S: They are right next to each other. → The most frequent measurement is in between the second and third most frequent measurements.

T: What do you notice about the data before the three most frequent measurements?

S: It goes 1, 2, 3, 3. → Hey, the number of plants goes up and then stays the same. → The number of plants increases or stays the same as it gets close to the most frequent measurement.

T: How about the data after the three most frequent measurements?

S: It goes 3, 2. → It starts to go back down! → After the most frequent measurement, the number of sunflower plants decreases for each measurement.

**MP.7**

T: (Cover up the bottom three rows of data in the chart.)  Erase the Xs on your line plot and create a new line plot with this data.  (Allow students time to work.)  Did the three most frequent measurements change when you plotted less data?

S: Yes, now the three most frequent measurements are 61, $61\frac{1}{2}$, and 62 inches.

T: That means that most of the sunflowers in Mrs. Schaut's garden are between 61 and 62 inches tall.

S: No, that is not right! → No, we saw earlier that most of the sunflowers are between 62 and 63 inches tall.

T: How did using less data change how we can talk about the heights of most of the sunflowers?  Discuss with your partner.

S: When we used less data, it changed the most frequent measurements. → Yeah, with more data we said most sunflowers were between 62 and 63 inches tall.  But with less data, that changed to between 61 and 62 inches.

T: How did the shape of the line plot change when we used less data?  Talk to a partner.

S:   The height of the line plot changed because with more data, the most X's for a measurement was 7, but with less data, the most X's is 3. → The three most frequent measurements shifted to the left on the number line. → It does not really follow the same pattern as increasing before the three most frequent measurements and decreasing after the three most frequent measurements. → Except for the three most frequent measurements, all other measurements only have one X.

**MP.7**

## Problem Set  (10 minutes)

Students should do their personal best to complete the Problem Set within the allotted 10 minutes.  For some classes, it may be appropriate to modify the assignment by specifying which problems they work on first.  Some problems do not specify a method for solving.  Students solve these problems using the RDW approach used for Application Problems.

## Student Debrief  (10 minutes)

**Lesson Objective:**  Represent measurement data with line plots.

The Student Debrief is intended to invite reflection and active processing of the total lesson experience.

Invite students to review their solutions for the Problem Set.  They should check work by comparing answers with a partner before going over answers as a class.  Look for misconceptions or misunderstandings that can be addressed in the Debrief.  Guide students in a conversation to debrief the Problem Set and process the lesson.

Any combination of the questions below may be used to lead the discussion.

- Look at Problem (b).  With a partner, compare the steps you took to create the line plot.

- (Invite students to share thinking for Problem (d).)  What can you say about most of the leaves from Delilah's tree?

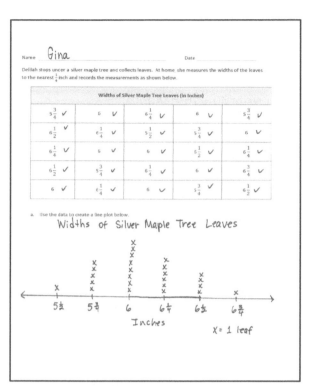

©2015 Great Minds. eureka-math.org
G3-M6-TE-B6-1.3.1-01.2016

- If the only **measurement data** we had was the top two rows of the chart, how might that change your understanding of the width of most of Delilah's leaves?

- Why does having a large amount of data help us have a clearer understanding of what the data means?

- Compare the shape of this data to that of the bean plants from yesterday. Why might the bean plants have grown so irregularly whereas the sunflower plants did not? Might their environments have been different?

- Looking at the size of most of the leaves from Delilah's tree, do you know any trees in your neighborhood that might be the same kind? Do you know any that are certainly not the same kind? (Students might talk about trees they see in the park or in their neighborhood, such as "the tree in the schoolyard," etc.)

## Exit Ticket  (3 minutes)

After the Student Debrief, instruct students to complete the Exit Ticket. A review of their work will help with assessing students' understanding of the concepts that were presented in today's lesson and planning more effectively for future lessons. The questions may be read aloud to the students.

Multiply.

7 x 1 = _____     7 x 2 = _____     7 x 3 = _____     7 x 4 = _____

7 x 5 = _____     7 x 1 = _____     7 x 2 = _____     7 x 1 = _____

7 x 3 = _____     7 x 1 = _____     7 x 4 = _____     7 x 1 = _____

7 x 5 = _____     7 x 1 = _____     7 x 2 = _____     7 x 3 = _____

7 x 2 = _____     7 x 4 = _____     7 x 2 = _____     7 x 5 = _____

7 x 2 = _____     7 x 1 = _____     7 x 2 = _____     7 x 3 = _____

7 x 1 = _____     7 x 3 = _____     7 x 2 = _____     7 x 3 = _____

7 x 4 = _____     7 x 3 = _____     7 x 5 = _____     7 x 3 = _____

7 x 4 = _____     7 x 1 = _____     7 x 4 = _____     7 x 2 = _____

7 x 4 = _____     7 x 3 = _____     7 x 4 = _____     7 x 5 = _____

7 x 4 = _____     7 x 5 = _____     7 x 1 = _____     7 x 5 = _____

7 x 2 = _____     7 x 5 = _____     7 x 3 = _____     7 x 5 = _____

7 x 4 = _____     7 x 2 = _____     7 x 4 = _____     7 x 3 = _____

7 x 5 = _____     7 x 3 = _____     7 x 2 = _____     7 x 4 = _____

7 x 3 = _____     7 x 5 = _____     7 x 2 = _____     7 x 4 = _____

multiply by 7 (1–5)

**Lesson 8:**     Represent measurement data with line plots.

EUREKA MATH™

Name _____ Date _____

Delilah stops under a silver maple tree and collects leaves. At home, she measures the widths of the leaves to the nearest $\frac{1}{4}$ inch and records the measurements as shown below.

| Widths of Silver Maple Tree Leaves (in Inches) | | | | |
|---|---|---|---|---|
| $5\frac{3}{4}$ | $6$ | $6\frac{1}{4}$ | $6$ | $5\frac{3}{4}$ |
| $6\frac{1}{2}$ | $6\frac{1}{4}$ | $5\frac{1}{2}$ | $5\frac{3}{4}$ | $6$ |
| $6\frac{1}{4}$ | $6$ | $6$ | $6\frac{1}{2}$ | $6\frac{1}{4}$ |
| $6\frac{1}{2}$ | $5\frac{3}{4}$ | $6\frac{1}{4}$ | $6$ | $6\frac{3}{4}$ |
| $6$ | $6\frac{1}{4}$ | $6$ | $5\frac{3}{4}$ | $6\frac{1}{2}$ |

a. Use the data to create a line plot below.

b. Explain the steps you took to create the line plot.

c. How many more leaves were 6 inches wide than $6\frac{1}{2}$ inches wide?

d. Find the three most frequent measurements on the line plot. What does this tell you about the typical width of a silver maple tree leaf?

EUREKA
MATH

Name _____     Date _____

The line plot below shows the lengths of fish the fishing boat caught.

**Lengths of Fish**

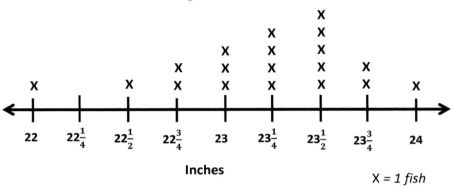

**Inches**

X = 1 fish

a.   Find the three most frequent measurements on the line plot.

b.   Find the difference between the lengths of the longest and shortest fish.

c.   How many more fish were $23\frac{1}{4}$ inches long than 24 inches long?

Name _____ Date _____

Mrs. Leah's class uses what they learned about simple machines to build marshmallow launchers. They record the distances their marshmallows travel in the chart below.

| Distance Traveled (in Inches) | | | | |
|---|---|---|---|---|
| $48\frac{3}{4}$ | 49 | $49\frac{1}{4}$ | 50 | $49\frac{3}{4}$ |
| $49\frac{1}{2}$ | $48\frac{1}{4}$ | $49\frac{1}{2}$ | $48\frac{3}{4}$ | 49 |
| $49\frac{1}{4}$ | $49\frac{3}{4}$ | 48 | $49\frac{1}{4}$ | $48\frac{1}{4}$ |
| 49 | $48\frac{3}{4}$ | 49 | 49 | $48\frac{3}{4}$ |

a. Use the data to create a line plot below.

**Lesson 8:** Represent measurement data with line plots.

EUREKA MATH™

b. Explain the steps you took to create the line plot.

c. How many more marshmallows traveled $48\frac{3}{4}$ inches than $48\frac{1}{4}$ inches?

d. Find the three most frequent measurements on the line plot. What does this tell you about the distance that most of the marshmallows traveled?

Mrs. Schaut measures the heights of the sunflower plants in her garden. The measurements are shown in the chart below.

| Heights of Sunflower Plants (in Inches) | | | | |
|---|---|---|---|---|
| 61 | 63 | 62 | 61 | $62\frac{1}{2}$ |
| $61\frac{1}{2}$ | $61\frac{1}{2}$ | $61\frac{1}{2}$ | 62 | 60 |
| 64 | 62 | $60\frac{1}{2}$ | $63\frac{1}{2}$ | 61 |
| 63 | $62\frac{1}{2}$ | $62\frac{1}{2}$ | 64 | $62\frac{1}{2}$ |
| $62\frac{1}{2}$ | $63\frac{1}{2}$ | 63 | $62\frac{1}{2}$ | $63\frac{1}{2}$ |
| 62 | $62\frac{1}{2}$ | 62 | 63 | $60\frac{1}{2}$ |

heights of sunflower plants chart

**Lesson 8:** Represent measurement data with line plots.

EUREKA
MATH

# Lesson 9

Objective:  Analyze data to problem solve.

## Suggested Lesson Structure

■ Fluency Practice          (14 minutes)
▨ Application Problem        (5 minutes)
☐ Concept Development        (31 minutes)
■ Student Debrief           (10 minutes)
    **Total Time**          **(60 minutes)**

## Fluency Practice  (14 minutes)

▪ Group Counting  **3.OA.1**                    (3 minutes)
▪ Multiply by 7  **3.OA.7**                      (7 minutes)
▪ Count by Halves and Fourths  **3.MD.4**        (4 minutes)

### Group Counting  (3 minutes)

Materials:   (S) Personal white board

Note:  This group counting activity reviews the relationship between counting by a unit and multiplying and dividing with that unit.

   T:   Count by sixes to 60.
   S:   6, 12, 18, 24, 30, 36, 42, 48, 54, 60.
   T:   (Write 4 sixes = ___.)  Write the number sentence.
   S:   (Write 4 sixes = 24.)
   T:   Write 4 sixes as a multiplication sentence.
   S:   (Write $4 \times 6 = 24$.)
   T:   (Write $48 \div 6$ = ___.)  Write the number sentence.  Count by sixes if you are unsure.
   S:   (Write $48 \div 6 = 8$.)
   T:   Count by eights to 80.
   S:   8, 16, 24, 32, 40, 48, 56, 64, 72, 80.
   T:   (Write 3 eights = ___.)  Write the number sentence.
   S:   (Write 3 eights = 24.)
   T:   Write 3 eights as a multiplication sentence.
   S:   (Write $3 \times 8 = 24$.)

---

T:   (Write 56 ÷ 8 = ___.)  Write the number sentence.  Count by eights if you are unsure.

S:   (Write 56 ÷ 8 = 7.)

T:   Count by nines to 90.

S:   9, 18, 27, 36, 45, 54, 63, 72, 81, 90.

T:   (Write 4 nines = ___.)  Write the number sentence.

S:   (Write 4 nines = 36.)

T:   Write 4 nines as a multiplication sentence.

S:   (Write 4 × 9 = 36.)

T:   (Write 54 ÷ 9 = ___.)  Write the number sentence.  Count by nines if you are unsure.

S:   (Write 54 ÷ 9 = 6.)

## Multiply by 7  (7 minutes)

Materials:   (S) Multiply by 7 (6–10) (Pattern Sheet)

Note:  This activity builds fluency with multiplication facts using units of 7.  It works toward students knowing from memory all products of two one-digit numbers.  See Lesson 6 for the directions for administration of a Multiply-By Pattern Sheet.

T:   (Write 6 × 7 = _____.)  Let's skip-count up by sevens to solve.  Watch as I raise a finger for each seven. (Raise a finger for each number to track the count.  Record the skip-count answers on the board.)

S:   7, 14, 21, 28, 35, 42.

T:   Let's skip-count down to find the answer, too.  Start at 70.  (Count down with fingers as students say numbers.)

S:   70, 63, 56, 49, 42.

Continue with the following suggested sequence:  8 × 7, 7 × 7, and 9 × 7.

T:   (Distribute the Multiply by 7 Pattern Sheet.)  Let's practice multiplying by 7.  Be sure to work left to right across the page.

## Count by Halves and Fourths  (4 minutes)

Note:  This activity reviews Lesson 6.

T:   Count by halves to 12 halves as I write.  Please do not count faster than I can write.  (Write as students count.)

S:   1 half, 2 halves, 3 halves, 4 halves, 5 halves, 6 halves, 7 halves, 8 halves, 9 halves, 10 halves, 11 halves, 12 halves.

T:   (Point to $\frac{2}{2}$.)  Say 2 halves as a whole number.

S:   1.

Halves:

$\frac{1}{2}$  $\frac{2}{2}$  $\frac{3}{2}$  $\frac{4}{2}$  $\frac{5}{2}$  $\frac{6}{2}$  $\frac{7}{2}$  $\frac{8}{2}$  $\frac{9}{2}$  $\frac{10}{2}$  $\frac{11}{2}$  $\frac{12}{2}$

1        2        3        4        5        6

Fourths:

$\frac{1}{4}$  $\frac{2}{4}$  $\frac{3}{4}$  $\frac{4}{4}$  $\frac{5}{4}$  $\frac{6}{4}$  $\frac{7}{4}$  $\frac{8}{4}$  $\frac{9}{4}$  $\frac{10}{4}$  $\frac{11}{4}$  $\frac{12}{4}$

1              2              3

T:    (Lightly cross out $\frac{2}{2}$ and write 1 beneath it.)

Continue the process for the following sequence: $\frac{4}{2}, \frac{6}{2}, \frac{8}{2}, \frac{10}{2}$, and $\frac{12}{2}$.

T:    Count by halves, saying whole numbers when you arrive at whole numbers.  Try not to look at the board.  (Direct students to count forward and backward on the number line, occasionally changing directions.)

Repeat the process for fourths.

## Application Problem  (5 minutes)

Marla creates a line plot with a half-inch scale from 33 to 37 inches.  How many tick marks should be on her line plot?

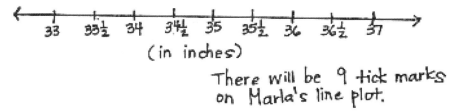

There will be 9 tick marks on Marla's line plot.

Note:  This problem reviews the concepts taught in Lessons 7–8.  Invite students to share their strategies for solving this problem.

## Concept Development  (31 minutes)

Materials:   (S) Bar graph and line plot (Template) shown below to the right, personal white board

**Problem 1:  Solve problems with categorical data.**

Project the bar graph from the Template as shown.

T:    This graph shows how some friends spent their money at the fair.

Project or read the following problem:  How much more money was spent on rides than on parking?

T:    How can you use the graph to help you solve this problem?  Talk to a partner.

S:    I can read the value of the rides bar and then subtract the value of the parking bar. $35 – $5.

T:    Choose a strategy and solve.  (Allow students time to work.)  How much more money was spent on rides than on parking?

S:    $30.

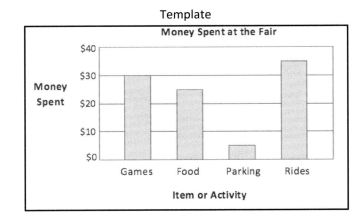

Template

T:   Talk to your partner:  Why do you think more money was spent on rides than on parking?

S:   The rides are one of the most fun activities at the fair. → Yeah, it would not make sense if parking cost more than the rides. → If parking was more expensive than the rides, people might not come to the fair at all!

Project or read the following problem:  The friends take a total of $120 to the fair.  How much do they have left after the fair?

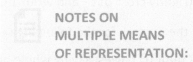

**NOTES ON MULTIPLE MEANS OF REPRESENTATION:**

Guide students to use tools within the graph to read values.  In Money Spent at the Fair, students should note that rows are at increments of $10. Keeping this in mind, guide students to quickly read the half unit as $5. Scaffold fluency by having students draw lines in each bar to make $10 units, connecting back to their reading of picture graphs.

T:   What is the first thing we need to find out?

S:   We need to find the total amount they spent at the fair.

T:   Talk to your partner.  How does the graph help us find the total amount?

S:   We can find how much the friends spent on each thing shown by the bar graph.  → Then, we can add the amounts together.

T:   Use the graph to write a number sentence to show how much money the friends spend in all.

S:   (Write $30 + $25 + $5 + $35 = $95.)

T:   How much do the friends spend in all?

S:   $95.

T:   Have we solved the problem?

S:   No.  We need to find how much money the friends have left.

T:   Write a number sentence to show how much money the friends have left.  (Allow students time to work.)  How much money do they have left after the fair?

S:   $25.

As time allows, continue with the additional questions below.  Students may work independently, in pairs, or in groups.

- How much less did the friends spend on rides than on games and food combined?
- Parking costs $1 for each hour.  The group of friends arrived at the fair at 3:00 p.m.  What time did they leave?

**Problem 2:  Solve problems with measurement data.**

Project the line plot from the Template as shown.

T:   This line plot shows the lengths of the crayfish in Mr. Nye's third-grade science class.

Project or read the following problem:  What is the total length of all the crayfish that are 3 inches long?

T:   Talk to your partner.  How can you use the line plot to help you solve this problem?

S:   I can skip-count the X's on the 3-inch mark by three. → I know there are 6 crayfish that are 3 inches long, so I can just multiply 6 times 3.

Template

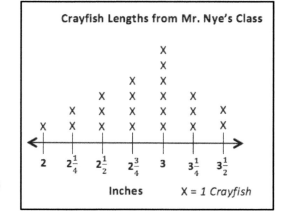

**EUREKA MATH**

T:  Solve. (Allow students time to work.)  What is the total length?

S:  18 inches!

Project or read the following problem:  Mrs. Curie's students also measure the lengths of their crayfish.  They notice the number of crayfish that are less than 3 inches long is half of the number of crayfish that are 3 inches long in Mr. Nye's class.  How many crayfish are less than 3 inches long in Mrs. Curie's class?

T:  What do you need to figure out first to solve this problem?

S:  The number of crayfish in Mr. Nye's class that are less than 3 inches long.

MP.2

T:  Discuss with a partner how to find the number of crayfish in Mr. Nye's class that are less than 3 inches long.

S:  (Discuss.)

T:  How many crayfish are less than 3 inches long in Mr. Nye's class?

S:  10 crayfish!

T:  How does this help you find the answer to the problem?

S:  Well, Mrs. Curie's class has half as many, so I can just divide 10 by 2. → I know that half of 10 is 5.

T:  How many crayfish are less than 3 inches long in Mrs. Curie's class?

S:  5 crayfish!

As time allows, continue with the additional questions below.  Students may work independently, in pairs, or in groups.

- Ginny uses half-inch square tiles to measure the longest crayfish.  How many half-inch square tiles does she use?

- Use the line plot and the chart below to find the total number of crayfish that all of the third-grade classes are studying.

| Classroom | Mr. Franklin | Mrs. Curie | Mr. Nye | Mrs. Nobel |
|---|---|---|---|---|
| Number of Crayfish | 21 | 23 | ? | 24 |

- The crayfish are kept in small tanks.  There are 3 crayfish in each tank.  How many tanks does Mr. Nye's class need?

T:  Data is shown in different forms depending on how it is used.  Compare the money spent at the fair problem to Mr. Nye's class's crayfish problem.  Talk to your partner.  Would it make sense for the money spent at the fair data to be switched to a line plot?  Explain why or why not.  Think about how each representation helps you analyze the data.

S:  Line plots usually show how many times a certain thing happens, like how many crayfish are a certain measurement.  It would not make sense to try to show money spent at the fair on a line plot. → We use a number line to make a line plot.  It would not make sense to put rides, food, games, and parking as labels on a number line! → What would each X represent?

T:  Bar graphs are used to compare things between different groups, and line plots are used to show frequency of data along a number line.

T:   Turn and talk to your partner.  If we wanted to show the number of coins in 4 piggy banks, what graph would you use and why?

S:   A bar graph, because we have 4 different groups.  → It does not make sense to plot piggy banks on a number line since we are comparing what is in each piggy bank.

If needed and time permits, continue asking students about which graph would be most appropriate for specific data.  The chart below shows some of the titles of bar graphs and line plots they have seen in this module.

| Bar Graphs | Line Plots |
|---|---|
| <ul><li>Number of fish in each tank</li><li>Number of students in each class</li><li>Amount of money saved each month</li><li>Number of magazines sold by each student</li><li>Number of visitors to a carnival each day</li><li>Number of coins in each piggy bank</li></ul> | <ul><li>Lengths of straws</li><li>Time spent outside over the weekend</li><li>Heights of children on a third-grade basketball team</li><li>Lengths of worms</li><li>Lengths of plants' roots</li><li>Heights of bean plants</li><li>Heights of sunflower plants</li><li>Widths of silver maple tree leaves</li></ul> |

## Problem Set  (10 minutes)

Students should do their personal best to complete the Problem Set within the allotted 10 minutes.  For some classes, it may be appropriate to modify the assignment by specifying which problems they work on first.  Some problems do not specify a method for solving.  Students should solve these problems using the RDW approach used for Application Problems.

NOTES ON
MULTIPLE MEANS
OF ENGAGEMENT:

Instead of completing the last word problem of the Problem Set, offer students working above grade level an open-ended challenge similar to the ones listed below:

- Represent the information from Lengths of Blades of Grass (in Inches) in a different type of graph. How does the presentation change your perception and understanding of the data?

- What other information might you obtain if you were to make a line plot for the Number of Apples Picked picture graph?

## Student Debrief  (10 minutes)

**Lesson Objective:** Analyze data to problem solve.

The Student Debrief is intended to invite reflection and active processing of the total lesson experience.

Invite students to review their solutions for the Problem Set.  They should check work by comparing answers with a partner before going over answers as a class.  Look for misconceptions or misunderstandings that can be addressed in the Debrief.  Guide students in a conversation to debrief the Problem Set and process the lesson.

Any combination of the questions below may be used to lead the discussion.

- What scale did you use for Problem 1(b)?  Would that scale work if Philip picked 21 apples?
- Compare your solution for Problem 2(b) to a partner's solution.  Did you and your partner use the same strategy to solve the problem?
- Explain to your partner how you chose the scale for the line plot in Problem 3(a).
- Other than counting the X's, is there another strategy you can use to find the total number of blades of grass that were measured in Problem 3(b)?  (Count the boxes in the chart, or multiply to find the total number of boxes in the chart.)
- Would it make sense to display the number of apples picked data in a line plot?  Why or why not?
- When is it best to show your data as a picture graph?  A bar graph?  A line plot?  What is the difference?

## Exit Ticket  (3 minutes)

After the Student Debrief, instruct students to complete the Exit Ticket.  A review of their work will help with assessing students' understanding of the concepts that were presented in today's lesson and planning more effectively for future lessons.  The questions may be read aloud to the students.

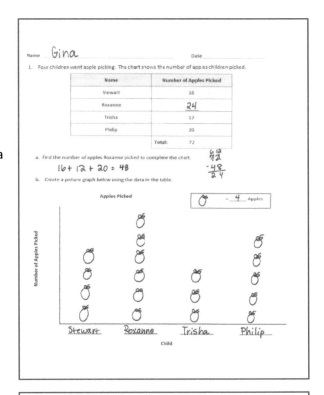

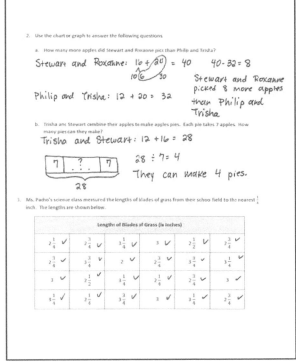

Lesson 9:        Analyze data to problem solve.                                                        **125**

©2015 Great Minds. eureka-math.org
G3-M6-TE-B6-1.3.1-01.2016

Multiply.

$7 \times 1 =$ _____     $7 \times 2 =$ _____     $7 \times 3 =$ _____     $7 \times 4 =$ _____

$7 \times 5 =$ _____     $7 \times 6 =$ _____     $7 \times 7 =$ _____     $7 \times 8 =$ _____

$7 \times 9 =$ _____     $7 \times 10 =$ _____     $7 \times 5 =$ _____     $7 \times 6 =$ _____

$7 \times 5 =$ _____     $7 \times 7 =$ _____     $7 \times 5 =$ _____     $7 \times 8 =$ _____

$7 \times 5 =$ _____     $7 \times 9 =$ _____     $7 \times 5 =$ _____     $7 \times 10 =$ _____

$7 \times 6 =$ _____     $7 \times 5 =$ _____     $7 \times 6 =$ _____     $7 \times 7 =$ _____

$7 \times 6 =$ _____     $7 \times 8 =$ _____     $7 \times 6 =$ _____     $7 \times 9 =$ _____

$7 \times 6 =$ _____     $7 \times 7 =$ _____     $7 \times 6 =$ _____     $7 \times 7 =$ _____

$7 \times 8 =$ _____     $7 \times 7 =$ _____     $7 \times 9 =$ _____     $7 \times 7 =$ _____

$7 \times 8 =$ _____     $7 \times 6 =$ _____     $7 \times 8 =$ _____     $7 \times 7 =$ _____

$7 \times 8 =$ _____     $7 \times 9 =$ _____     $7 \times 9 =$ _____     $7 \times 6 =$ _____

$7 \times 9 =$ _____     $7 \times 7 =$ _____     $7 \times 9 =$ _____     $7 \times 8 =$ _____

$7 \times 9 =$ _____     $7 \times 8 =$ _____     $7 \times 6 =$ _____     $7 \times 9 =$ _____

$7 \times 7 =$ _____     $7 \times 9 =$ _____     $7 \times 6 =$ _____     $7 \times 8 =$ _____

$7 \times 9 =$ _____     $7 \times 7 =$ _____     $7 \times 6 =$ _____     $7 \times 8 =$ _____

multiply by 7 (6–10)

**Lesson 9:**     Analyze data to problem solve.

EUREKA MATH™

Name _____ Date _____

1.  Four children went apple picking. The chart shows the number of apples the children picked.

| Name | Number of Apples Picked |
|------|------------------------|
| Stewart | 16 |
| Roxanne | _____ |
| Trisha | 12 |
| Philip | 20 |
| **Total:** | 72 |

a.  Find the number of apples Roxanne picked to complete the chart.


b.  Create a picture graph below using the data in the table.


**Apples Picked**

| = _____ Apples |

Number of Apples Picked

_____    _____    _____    _____

**Child**

2.  Use the chart or graph to answer the following questions.

    a.  How many more apples did Stewart and Roxanne pick than Philip and Trisha?

    b.  Trisha and Stewart combine their apples to make apples pies.  Each pie takes 7 apples.  How many pies can they make?

3.  Ms. Pacho's science class measured the lengths of blades of grass from their school field to the nearest $\frac{1}{4}$ inch.  The lengths are shown below.

| Lengths of Blades of Grass (in Inches) | | | | | |
|---|---|---|---|---|---|
| $2\frac{1}{4}$ | $2\frac{3}{4}$ | $3\frac{1}{4}$ | $3$ | $2\frac{1}{2}$ | $2\frac{3}{4}$ |
| $2\frac{3}{4}$ | $3\frac{3}{4}$ | $2$ | $2\frac{3}{4}$ | $3\frac{3}{4}$ | $3\frac{1}{4}$ |
| $3$ | $2\frac{1}{2}$ | $3\frac{1}{4}$ | $2\frac{1}{4}$ | $2\frac{3}{4}$ | $3$ |
| $3\frac{1}{4}$ | $2\frac{1}{4}$ | $3\frac{3}{4}$ | $3$ | $3\frac{1}{4}$ | $2\frac{3}{4}$ |

EUREKA
MATH™

©2015 Great Minds. eureka-math.org
G3-M6-TE-B6-1.3.1-01.2016

a.  Make a line plot of the grass data.  Explain your choice of scale.

b.  How many blades of grass were measured?  Explain how you know.

c.  What was the length measured most frequently on the line plot?  How many blades of grass had this length?

d.  How many more blades of grass measured $2\frac{3}{4}$ inches than both $3\frac{3}{4}$ inches and 2 inches combined?

Name _____     Date _____

Mr. Gallagher's science class goes bird watching.  The picture graph below shows the number of birds the class observes.

 = ___6___ Birds

**Number of Birds Mr. Gallagher's Class Observed**

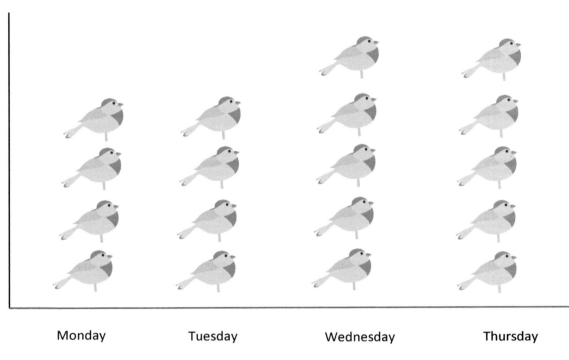

a.  How many more birds did Mr. Gallagher's class observe on Wednesday and Thursday than on Monday and Tuesday?

b.  Mr. Manning's class observed 104 birds.  How many more birds did Mr. Gallagher's class observe?

Name _____   Date _____

1.  The table below shows the amount of money Danielle saves for four months.

| Month | Money Saved |
|---|---|
| January | $9 |
| February | $18 |
| March | $36 |
| April | $27 |

Create a picture graph below using the data in the table.

**Money Danielle Saves**

= _____ Dollars

Money Saved

_____   _____   _____   _____

**Month**

2. Use the table or graph to answer the following questions.

   a. How much money does Danielle save in four months?

   b. How much more money does Danielle save in March and April than in January and February?

   c. Danielle combines her savings from March and April to buy books for her friends. Each book costs $9. How many books can she buy?

   d. Danielle earns $33 in June. She buys a necklace for $8 and a birthday present for her brother. She saves the $13 she has left. How much does the birthday present cost?

Lesson 9: Analyze data to problem solve.

©2015 Great Minds. eureka-math.org
G3-M6-TE-B6-1.3.1-01.2016

EUREKA MATH

**Money Spent at the Fair**

**Crayfish Lengths from Mr. Nye's Class**

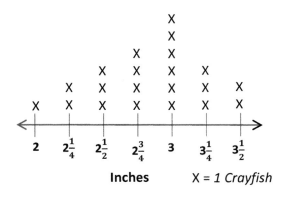

bar graph and line plot

Name _____    Date _____

1.  The picture graph below represents all the trees in the park.

**Trees in the Park**

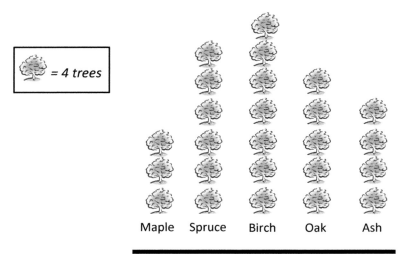

= 4 trees

Maple    Spruce    Birch    Oak    Ash

**Type of Tree**

a.  Use the grid to create and label a scaled bar graph representing the data in the picture graph above.

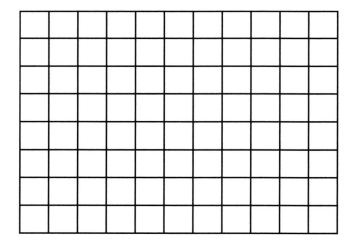

b.  How many more maple and oak trees are there than birch trees?

EUREKA
MATH™

2. The table below shows the number of flowers that were planted by the science club.

   a. Complete the table by filling in the number of marigolds that were planted.

   | Flowers Planted by Science Club | |
   | --- | --- |
   | Type of Flower | Number Planted |
   | Roses | 24 |
   | Lilies | 12 |
   | Marigolds | _____ |
   | TOTAL Flowers Planted: | 54 |

   b. Use the lines below to create and label a picture graph using the data in the table. Determine a picture and scale to represent the number of each type of flower.

   |  |
   | --- |
   | = _____ *flowers* |

   **Number Planted**

   _____  _____  _____

   **Type of Flower**

3. Fred measures the heights of all the sunflowers in his backyard.  His measurements in inches are shown on the line plot below.

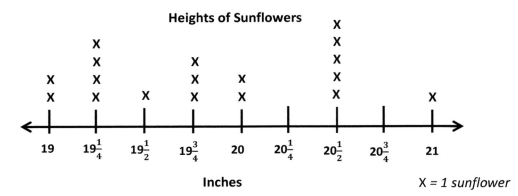

**Heights of Sunflowers**

**Inches**                                    X = 1 sunflower

a.  How many sunflowers are in Fred's backyard?  Explain how you know.

b.  What are the three most frequent measurements on the line plot?  Write them in order from shortest to longest.

©2015 Great Minds. eureka-math.org
G3-M6-TE-B6-1.3.1-01.2016

EUREKA
MATH™

4. Carol measures 16 bamboo shoots. Her measurements are recorded in the table below.

| Heights of Bamboo Shoots (in Inches) | | | |
|---|---|---|---|
| $94\frac{1}{2}$ | $94\frac{1}{4}$ | $93\frac{3}{4}$ | $94\frac{3}{4}$ |
| $94\frac{3}{4}$ | $95$ | $94\frac{3}{4}$ | $95\frac{1}{4}$ |
| $94\frac{1}{2}$ | $94\frac{3}{4}$ | $94\frac{3}{4}$ | $94\frac{1}{2}$ |
| $95$ | $94\frac{3}{4}$ | $94\frac{3}{4}$ | $95$ |

a. Make a line plot of the bamboo shoot data. Explain your choice of scale.

b. How many more bamboo shoots measured $94\frac{3}{4}$ inches than both 95 and $94\frac{1}{2}$ inches combined?

G3-M6-TE-B6-1.3.1-01.2016

| End-of-Module Assessment Task | Topics A–B |
| --- | --- |
| Standards Addressed | |

**Represent and interpret data.**

**3MD.3**  Draw a scaled picture graph and a scaled bar graph to represent a data set with several categories.  Solve one- and two-step "how many more" and "how many less" problems using information presented in scaled bar graphs.  *For example, draw a bar graph in which each square in the bar graph might represent 5 pets.*

**3.MD.4**  Generate measurement data by measuring lengths using rulers marked with halves and fourths of an inch.  Show the data by making a line plot, where the horizontal scale is marked off in appropriate units—whole numbers, halves, or quarters.

## Evaluating Student Learning Outcomes

A Progression Toward Mastery is provided to describe steps that illuminate the gradually increasing understandings that students develop on their way to proficiency.  In this chart, this progress is presented from left (Step 1) to right (Step 4).  The learning goal for each student is to achieve Step 4 mastery.  These steps are meant to help teachers and students identify and celebrate what the students CAN do now and what they need to work on next.

## A Progression Toward Mastery

| Assessment Task Item and Standards Assessed | STEP 1 Little evidence of reasoning without a correct answer. (1 Point) | STEP 2 Evidence of some reasoning without a correct answer. (2 Points) | STEP 3 Evidence of some reasoning with a correct answer or evidence of solid reasoning with an incorrect answer. (3 Points) | STEP 4 Evidence of solid reasoning with a correct answer. (4 Points) |
|---|---|---|---|---|
| **1**<br><br>3.MD.3 | Student is unable to answer either question correctly. | Student attempts to draw and label the bar graph but does not use an appropriate scale. Student may or may not find the correct answer in part (b). | Student creates an accurate bar graph with labels and an appropriate scale but does not answer part (b) correctly.<br>OR<br>Student creates a graph that is missing labels but is otherwise correct, and part (b) is correct. | Student:<br>a. Creates an accurate, labeled bar graph with a scale of 4.<br>b. Finds that there are 4 more maple and oak trees than birch trees. |
| **2**<br><br>3.MD.3 | Student attempts to complete the table but finds the incorrect number of marigolds in part (a). Student is unable to correctly complete the picture graph in part (b). | Student correctly calculates the number of marigolds and attempts to scale, create, and label a picture graph in part (b). | Student correctly calculates 18 marigolds in part (a) and correctly scales and labels the picture graph in part (b) but incorrectly represents the number of flowers for one or more types in part (b). | Student correctly:<br>a. Calculates 18 marigolds.<br>b. Determines an appropriate scale and graphic representation; creates an accurate, labeled picture graph based on the data in the table. |
| **3**<br><br>3.MD.4 | Student is unable to answer any question correctly. | Student correctly answers either part (a) or part (b). | Student correctly answers part (a) and correctly identifies $19\frac{1}{4}$, $19\frac{3}{4}$, and $20\frac{1}{2}$ in Part (b) but does not list the measurements in order. | Student:<br>a. Finds 18 sunflowers in Fred's backyard and provides sound reasoning to support the answer.<br>b. Lists $19\frac{1}{4}$, $19\frac{3}{4}$, and $20\frac{1}{2}$ in order. |

©2015 Great Minds. eureka-math.org
G3-M6-TE-B6-1.3.1-01.2016

| A Progression Toward Mastery | | | | |
|---|---|---|---|---|
| **4**<br><br>**3.MD.4** | Student attempts, but is unable to complete, either question correctly. | Student draws the line plot correctly but may not explain her choice of scale. Student may make a minor error calculating part (b). | Student answers part (a) correctly but makes a minor error calculating part (b). | Student:<br><br>a. Creates an appropriate scale, draws a line plot to accurately display the data, and provides sound reasoning for the choice of scale.<br><br>b. Finds 1 more shoot that measured $94\frac{3}{4}$ inches than 95 and $94\frac{1}{2}$ inches combined. |

Module 6:     Collecting and Displaying Data

**EUREKA MATH**

Name ___Gina_____     Date _____

1. The picture graph below represents all the trees in the park.

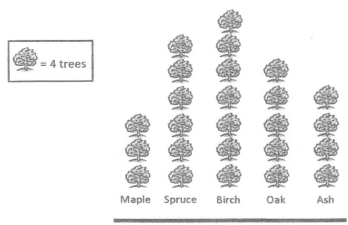

a. Use the grid to create and label a scaled bar graph representing the data in the picture graph above.

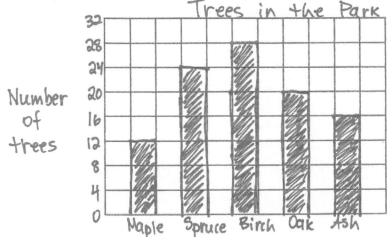

b. How many more maple and oak trees are there than birch trees?

32 Maple and Oak

$12 + 20 = 32$

28 Birch   ?    $32 - 28 = 4$

2 (30) 2

There are 4 more Maple and Oak than Birch trees.

2. The table below shows the number of flowers that were planted by the science club.

   a. Complete the table by filling in the number of marigolds that were planted.

| Flowers Planted by Science Club | |
| --- | --- |
| Type of Flower | Number Planted |
| Roses | 24 |
| Lilies | 12 |
| Marigolds | 18 |
| TOTAL Flowers Planted: | 54 |

$$> 54 - 36 = 18$$

14 40 4

   b. Use the lines below to create and label a picture graph using the data in the table. Determine a picture and scale to represent the number of each type of flower.

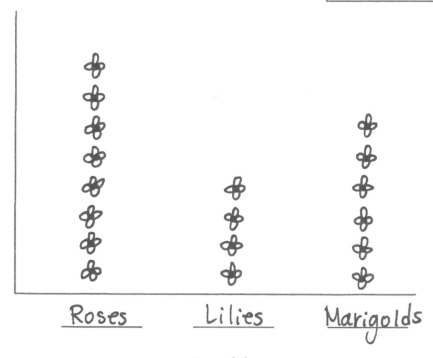

❀ = __3__ flowers

EUREKA
MATH™

3. Fred measures the heights of all the sunflowers in his backyard. His measurements in inches are shown on the line plot below.

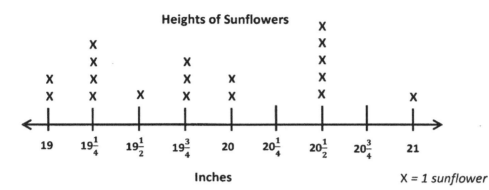

**Heights of Sunflowers**

Inches

X = 1 sunflower

a. How many sunflowers are in Fred's backyard? Explain how you know.

> There are 18 sunflowers in Fred's backyard.
> Each X represents 1 sunflower, so I found the
> total number of sunflowers by counting all of
> the X's.

b. What are the three most frequent measurements on the line plot? Write them in order from shortest, to longest.

> The 3 most frequent measurements on the
> line plot from shortest to longest are
> $19\frac{1}{4}$ inches, $19\frac{3}{4}$ inches, and $20\frac{1}{2}$ inches.

4. Carol measures 16 bamboo shoots. Her measurements are recorded in the table below.

| Heights of Bamboo Shoots (in Inches) | | | |
|---|---|---|---|
| $94\frac{1}{2}$ ✓ | $94\frac{1}{4}$ ✓ | $93\frac{3}{4}$ ✓ | $94\frac{3}{4}$ ✓ |
| $94\frac{3}{4}$ ✓ | $95$ ✓ | $94\frac{3}{4}$ ✓ | $95\frac{1}{4}$ ✓ |
| $94\frac{1}{2}$ ✓ | $94\frac{3}{4}$ ✓ | $94\frac{3}{4}$ ✓ | $94\frac{1}{2}$ ✓ |
| $95$ ✓ | $94\frac{3}{4}$ ✓ | $94\frac{3}{4}$ ✓ | $95$ ✓ |

a. Make a line plot of the bamboo shoot data. Explain your choice of scale.

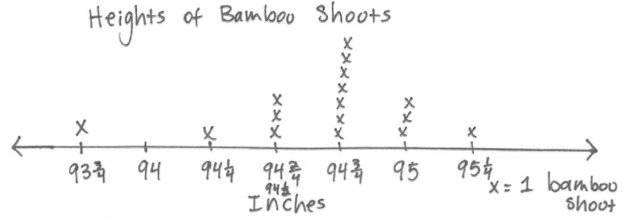

Heights of Bamboo Shoots

Inches

X = 1 bamboo shoot

I used fourths of an inch for my scale because I looked at all of the heights and saw that fourths of an inch was the smallest unit.

b. How many more bamboo shoots measured $94\frac{3}{4}$ inches than both 95 and $94\frac{1}{2}$ inches combined?

$94\frac{3}{4}$: 7 shoots

$95 + 94\frac{1}{2}$: $3 + 3 = 6$

1 more bamboo shoot measured $94\frac{3}{4}$ inches than 95 and $94\frac{1}{2}$ inches combined.

EUREKA MATH™

# Answer Key

# Eureka Math
# Grade 3
# Module 6

Special thanks go to the Gordon A. Cain Center and to the Department of Mathematics at Louisiana State University for their support in the development of *Eureka Math*.

For a free *Eureka Math* Teacher Resource Pack, Parent Tip Sheets, and more please visit www.Eureka.tools

**3**
GRADE

# Mathematics Curriculum

Answer Key
# GRADE 3 • MODULE 6

Collecting and Displaying Data

©2015 Great Minds. eureka-math.org
G3-M6-TE-B6-1.3.1-01.2016

# Lesson 1

## Problem Set

1. Tally chart will vary.

2. Answers will vary.

3. Picture graphs will vary.

4. a. 2 students

   b. 1 and a half hearts drawn correctly; 2 + 1 = 3

   c. 14 students; 7 × 2 = 14

   d. Answers will vary.

## Exit Ticket

a. Football; hockey

b. 9

c. 3; 12 − 9 = 3

d. 33

## Homework

1. 22

2. Answer provided; 4 circles; 2 circles; 8 circles; 2 circles

   a. Fish; lizards

   b. 8

   c. 2

3. 3 rectangles; 2 rectangles; 1 rectangle; 4 rectangles; 1 rectangle

   a. 2 students

   b. 10 students; 5 × 2 = 10

   c. 3; 8 − 2 = 6

EUREKA MATH™

# Lesson 2

## Problem Set

1.  Answer provided; 2 units of 4 drawn; 6 units of 4 drawn; 8 units of 4 drawn

2.  Answers will vary.

3.  a.  Answer provided; 2 units of 4 drawn; 6 units of 4 drawn; 8 units of 4 drawn

    b.  Answer provided; 1 unit of 8 drawn; 3 units of 8 drawn; 4 units of 8 drawn

    c.  Answers will vary.

    d.  20

    e.  10

    f.  Answers will vary.

    g.  Answers will vary.

## Exit Ticket

a.  3 units of 4 drawn; 4 units of 4 drawn; 5 units of 4 drawn; 2 units of 4 drawn

b.  Answers will vary.

## Homework

1.  Answer provided; 8 units of 2 drawn; 6 units of 2 drawn; 2 units of 2 drawn

2.  Answers will vary.

3.  a.  Answer provided; 8 units of 2 drawn; 6 units of 2 drawn; 2 units of 2 drawn

    b.  Answer provided; 4 units of 4 drawn; 3 units of 4 drawn; 1 unit of 4 drawn

    c.  Answers will vary.

    d.  Answers will vary.

    e.  $6 \times 2 = 12$

    f.  $3 \times 4 = 12$

    g.  Number and size of units; explanations will vary.

# Lesson 3

## Sprint

### Side A

| | | | | | | | |
|---|---|---|---|---|---|---|---|
| 1. | 12 | 12. | 42 | 23. | 10 | 34. | 8 |
| 2. | 18 | 13. | 48 | 24. | 2 | 35. | 7 |
| 3. | 24 | 14. | 54 | 25. | 3 | 36. | 9 |
| 4. | 30 | 15. | 60 | 26. | 10 | 37. | 6 |
| 5. | 6 | 16. | 8 | 27. | 5 | 38. | 8 |
| 6. | 2 | 17. | 7 | 28. | 1 | 39. | 66 |
| 7. | 3 | 18. | 9 | 29. | 2 | 40. | 11 |
| 8. | 5 | 19. | 6 | 30. | 3 | 41. | 72 |
| 9. | 1 | 20. | 10 | 31. | 6 | 42. | 12 |
| 10. | 4 | 21. | 5 | 32. | 7 | 43. | 84 |
| 11. | 36 | 22. | 1 | 33. | 9 | 44. | 14 |

### Side B

| | | | | | | | |
|---|---|---|---|---|---|---|---|
| 1. | 6 | 12. | 36 | 23. | 2 | 34. | 7 |
| 2. | 12 | 13. | 42 | 24. | 10 | 35. | 8 |
| 3. | 18 | 14. | 48 | 25. | 3 | 36. | 9 |
| 4. | 24 | 15. | 54 | 26. | 2 | 37. | 6 |
| 5. | 30 | 16. | 7 | 27. | 1 | 38. | 7 |
| 6. | 3 | 17. | 6 | 28. | 10 | 39. | 66 |
| 7. | 2 | 18. | 8 | 29. | 5 | 40. | 11 |
| 8. | 4 | 19. | 10 | 30. | 3 | 41. | 72 |
| 9. | 1 | 20. | 9 | 31. | 3 | 42. | 12 |
| 10. | 5 | 21. | 1 | 32. | 4 | 43. | 78 |
| 11. | 60 | 22. | 5 | 33. | 9 | 44. | 13 |

EUREKA MATH™

## Problem Set

1. Answer provided; 8 units colored; 6 and a half units colored; 9 units colored

   a. 2 students

   b. $9 + 16 + 13 + 18 = 56$

   c. $6$; $22 - 16 = 6$

2. a. $34

   b. February, April, and May

   c. $17$; $40 - $23 = $17$

   d. April; March

3. February: $30; March: $46; April: $23; May: $34; June: $40

4. Intervals drawn correctly on the number line; each day plotted and labeled correctly on number line

5. a. Monday, Tuesday, and Thursday; 50 min

   b. 30 min

## Exit Ticket

a. Intervals drawn correctly on the number line; each flavor plotted and labeled correctly on the number line

b. $25 + 35 + 50 = 110$

## Homework

1. a. 14

   b. 4; $18 - 14 = 4$

   c. $18 + 13 = 31$ and $17 + 14 = 31$; They get the same number of votes.

2. a. Week 4; Week 3

   b. 15 liters

   c. $50\ L + 40\ L = 90\ L$

   d. 210 liters

   e. $60 + 60 + 60 + 60 = 240$; More

3. Week 1: 55 liters; Week 2: 50 liters; Week 3: 40 liters; Week 4: 65 liters

EUREKA
MATH™

Module 6:     Collecting and Displaying Data

149

©2015 Great Minds. eureka-math.org
G3-M6-TE-B6-1.3.1-01.2016

# Lesson 4

## Problem Set

1. a. Bar graph drawn correctly with appropriate scale
   b. Answers will vary.
   c. 150
   d. 150
2. a. 240
   b. 80

## Exit Ticket

a. 530
b. 150

## Homework

1. 68; 62; 57; 24
   a. Bar graph drawn correctly with scale labeled appropriately by 10
   b. 11
   c. 171; work shown correctly
2. a. 5
   b. $118
   c. 11

EUREKA
MATH™

# Lesson 5

## Problem Set

1. a. Answers will vary.

   b. Answers will vary.

   c. Answers will vary.

2. a. Whole and half inches labeled on paper strip

   b. $\frac{1}{4}$ inch marks are drawn; 2; 4; 2

   c. Answers will vary.

3. Explanations will vary.

## Exit Ticket

a. Whole and quarter inches labeled on paper strip

b. Answers will vary.

## Homework

1. a. Red pencil; $6\frac{3}{4}$

   b. Green pencil; explanations will vary.

2. a. Whole and half inches labeled on paper strip

   b. 2; 4; 2; 1

3. Explanations will vary.

# Lesson 6

## Pattern Sheet

| | | | |
|---|---|---|---|
| 6 | 12 | 18 | 24 |
| 30 | 6 | 12 | 6 |
| 18 | 6 | 24 | 6 |
| 30 | 6 | 12 | 18 |
| 12 | 24 | 12 | 30 |
| 12 | 6 | 12 | 18 |
| 6 | 18 | 12 | 18 |
| 24 | 18 | 30 | 18 |
| 24 | 6 | 24 | 12 |
| 24 | 18 | 24 | 30 |
| 24 | 30 | 6 | 30 |
| 12 | 30 | 18 | 30 |
| 24 | 12 | 24 | 18 |
| 30 | 18 | 12 | 24 |
| 18 | 30 | 12 | 24 |

## Problem Set

1.  a.   15, explanations will vary.

    b.   6

    c.   No, explanations will vary.

    d.   4

2.  a.   30, explanations will vary.

    b.   No, explanations will vary.

    c.   Length plotted correctly on line plot

EUREKA MATH

**Exit Ticket**

a. 24, explanations will vary.

b. 9

c. Yes, explanations will vary.

**Homework**

1. a. 4

   b. 20; explanations will vary.

   c. 9

   d. 9

2. a. 24, explanations will vary.

   b. No, explanations will vary.

   c. Yes, explanations will vary.

# Lesson 7

## Pattern Sheet

| | | | |
|---|---|---|---|
| 6 | 12 | 18 | 24 |
| 30 | 36 | 42 | 48 |
| 54 | 60 | 30 | 36 |
| 30 | 42 | 30 | 48 |
| 30 | 54 | 30 | 60 |
| 36 | 30 | 36 | 42 |
| 36 | 48 | 36 | 54 |
| 36 | 42 | 36 | 42 |
| 48 | 42 | 54 | 42 |
| 48 | 36 | 48 | 42 |
| 48 | 54 | 54 | 36 |
| 54 | 42 | 54 | 48 |
| 54 | 48 | 36 | 54 |
| 42 | 54 | 36 | 48 |
| 54 | 42 | 36 | 48 |

## Problem Set

a. Line plot completed; Heights of Bean Plants; inches; values for X may vary

b. 14

c. 6

d. $1\frac{3}{4}$ inches; 4

e. No; explanations will vary.

f. Yes; explanations will vary.

## Exit Ticket

Line plot completed; Length of Mice; inches

Module 6:    Collecting and Displaying Data

©2015 Great Minds. eureka-math.org
G3-M6-TE-B6-1.3.1-01.2016

## Homework

a.  Line plot completed; Heights of Buildings; inches; values for X may vary

b.  2

c.  7

d.  25; explanations will vary.

e.  No; explanations will vary.

# Lesson 8

## Pattern Sheet

| | | | |
|---|---|---|---|
| 7 | 14 | 21 | 28 |
| 35 | 7 | 14 | 7 |
| 21 | 7 | 28 | 7 |
| 35 | 7 | 14 | 21 |
| 14 | 28 | 14 | 35 |
| 14 | 7 | 14 | 21 |
| 7 | 21 | 14 | 21 |
| 28 | 21 | 35 | 21 |
| 28 | 7 | 28 | 14 |
| 28 | 21 | 28 | 35 |
| 28 | 35 | 7 | 35 |
| 14 | 35 | 21 | 35 |
| 28 | 14 | 28 | 21 |
| 35 | 21 | 14 | 28 |
| 21 | 35 | 14 | 28 |

## Problem Set

a.  Line plot completed; Widths of Leaves; inches; values for X may vary

b.  Answers will vary.

c.  4

d.  Explanations will vary.

## Exit Ticket

a.  23 inches, $23\frac{1}{4}$ inches, $23\frac{1}{2}$ inches

b.  2 inches

c.  3

EUREKA MATH

## Homework

a.   Line plot completed; Distance Traveled; inches; values for X may vary

b.   Answers will vary.

c.   2

d.   Explanations will vary.

# Lesson 9

## Pattern Sheet

| | | | |
|---|---|---|---|
| 7 | 14 | 21 | 28 |
| 35 | 42 | 49 | 56 |
| 63 | 70 | 35 | 42 |
| 35 | 49 | 35 | 56 |
| 35 | 63 | 35 | 70 |
| 42 | 35 | 42 | 49 |
| 42 | 56 | 42 | 63 |
| 42 | 49 | 42 | 49 |
| 56 | 49 | 63 | 49 |
| 56 | 42 | 56 | 49 |
| 56 | 63 | 63 | 42 |
| 63 | 49 | 63 | 56 |
| 63 | 56 | 42 | 63 |
| 49 | 63 | 42 | 56 |
| 63 | 49 | 42 | 56 |

## Problem Set

1.  a.  24

    b.  Picture graph completed; scale drawn

2.  a.  8

    b.  4

3.  a.  Line plot completed; explanations will vary.

    b.  24; explanations will vary.

    c.  $2\frac{3}{4}$ inches; 6

    d.  2

EUREKA MATH™

## Exit Ticket

a.   12

b.   4

## Homework

1.   Picture graph completed; scale drawn

2.   a.   $90

    b.   $36

    c.   7

    d.   $12

This page  intentionally left  blank